Applied Mathematical Sciences

EDITORS

Fritz John
Courant Institute of
Mathematical Sciences
New York University
New York, N.Y. 10012

Joseph P. LaSalle
Division of
Applied Mathematics
Brown University
Providence, R.I. 02912

Lawrence Sirovich
Division of
Applied Mathematics
Brown University
Providence, R.I. 02912

Gerald B. Whitham
Applied Mathematics
Firestone Laboratory
California Institute of Technology
Pasadena, CA. 91125

EDITORIAL STATEMENT

The mathematization of all sciences, the fading of traditional scientific boundaries, the impact of computer technology, the growing importance of mathematical-computer modelling and the necessity of scientific planning all create the need both in education and research for books that are introductory to and abreast of these developments.

The purpose of this series is to provide such books, suitable for the user of mathematics, the mathematician interested in applications, and the student scientist. In particular, this series will provide an outlet for material less formally presented and more anticipatory of needs than finished texts or monographs, yet of immediate interest because of the novelty of its treatment of an application or of mathematics being applied or lying close to applications.

The aim of the series is, through rapid publication in an attractive but inexpensive format, to make material of current interest widely accessible. This implies the absence of excessive generality and abstraction, and unrealistic idealization, but with quality of exposition as a goal.

Many of the books will originate out of and will stimulate the development of new undergraduate and graduate courses in the applications of mathematics. Some of the books will present introductions to new areas of research, new applications and act as signposts for new directions in the mathematical sciences. This series will often serve as an intermediate stage of the publication of material which, through exposure here, will be further developed and refined. These will appear in conventional format and in hard cover.

MANUSCRIPTS

The Editors welcome all inquiries regarding the submission of manuscripts for the series. Final preparation of all manuscripts will take place in the editorial offices of the series in the Division of Applied Mathematics, Brown University, Providence, Rhode Island.

SPRINGER-VERLAG NEW YORK INC., 175 Fifth Avenue, New York, N.Y. 10010

Printed in U.S.A.

Applied Mathematical Sciences | Volume 28

Julian Keilson

Markov Chain Models—Rarity and Exponentiality

Springer-Verlag New York Heidelberg Berlin

Julian Keilson
The University of Rochester
Rochester, New York 14627
USA

AMS Subject Classification: 60 J 10

Library of Congress Cataloging in Publication Data

Keilson, Julian.
Markov chain models—rarity and exponentiality.

(Applied mathematical sciences ; v. 28)
Bibliography: p.
Includes index.
1. Markov processes. I. Title. II. Series.
QA1.A647 vol. 28 [QA274.7] 510'.8s [519.2'33] 79-10967

Printed in the United States of America.

9 8 7 6 5 4 3 2 1

ISBN 0-387-90405-0 Springer-Verlag New York Heidelberg Berlin
ISBN 3-540-90405-0 Springer-Verlag Berlin Heidelberg New York

Foreword and Acknowledgment

The work here offered builds on ideas and results developed over the last ten years. These ideas have been motivated by questions arising in the study of neuron firing, human metabolism, congestion, inventories and system reliability. A series of informal lectures on these ideas was presented during the academic year 1973-74 while visiting the Department of Statistics at Stanford University and the Department of Mathematics at the University of Helsinki. The lectures were developed further during a visit to the Operations Research Center at Berkeley in the summer of 1974 into a somewhat tidier and more extensive form, and subsequent changes have been minor.

The object of the effort is the development and presentation of the working tools needed to quantify the ergodic and transient behavior of systems of many degrees of freedom such as arise, for example, in the study of system reliability. The structure of passage time densities and exit time densities to and from subsets of the state space is discussed at length, with special emphasis on the exponentiality and related structural properties present in these densities when the sets visited are seen infrequently. The underlying themes of reversibility in time and complete monotonicity are of particular importance. They are at the heart of much of the tractability. It is assumed that the reader is familiar with the elements of probability theory and stochastic processes as found, for example, in Feller, Vol. I, and parts of Feller, Vol. II. By and large, the material is self-contained.

The effort needed to present the material would not have been made without the encouragement of Professor R. E.

Barlow of Berkeley, whose infectious enthusiasm and stimulating discussions are gratefully acknowledged.

The labor involved was supported in large part by the technical and editorial effort of many associates and, in particular, of M. Heikkila, A. Kester, U. Sumita, and Y. Hayashi. Their help has been essential to the end product. The editorial assistance of L. Ziegenfuss in the final preparation of the manuscript is gratefully acknowledged.

Appreciation is expressed for the contributions of Professor S. C. Graves of M.I.T., Professor D. R. Smith of Columbia University, and Professor E. Arjas of Oulu. Thanks are also tended to Professor M. Brown of the City University of New York, Professor R. Syski of Maryland, and Professor W. Whitt of Yale University, for helpful discussions in recent years. Finally, I would like to express my appreciation to the Office of Naval Research for its support, direct and indirect, of this work.

Notation

$\underline{a}$, $\underline{b}$, $\underline{p}$, etc.	column vectors
$\underline{a}^T$, $\underline{b}^T$, $\underline{p}^T$, etc.	row vectors
$\underline{1}$	column vector with all components 1
$\mathbf{a}$, $\mathbf{b}$, $\mathbf{p}$, etc.	square matrices
$\mathbf{a}^T$, $\mathbf{b}^T$, $\mathbf{p}^T$, etc.	transposed matrices
$\mathbf{I}$	identity matrix
$\mathscr{N}$	state space
N_K	Markov chain in discrete time
$\mathbf{a}$	single-step transition probability matrix for N_k
$a_{mn}^{(k)} = [\mathbf{a}^k]_{mn}$	k-step transition probability from m to n
$\underline{p}_k$	state probability vector after k steps
$N(t)$	Markov chain in continuous time
$\boldsymbol{\nu} = [\nu_{mn}]$	transition rate matrix for N(t) where ν_{mn} is the transition rate from m to n
$\mathbf{p}(\tau)$	transition probability matrix for N(t) for elapsed time τ
$\underline{p}(t)$	state probability vector for N(t) at t
$\underline{e}$	ergodic probability vector
$\mathbf{e}_D = \text{diag}(e_n)$	diagonal matrix with the n-th diagonal component e_n of $\underline{e}$
$\mathbf{e}_D^{1/2} = \text{diag}(e_n^{1/2})$	diagonal matrix with the n-th diagonal component $e_n^{1/2}$
$\underline{a}\underline{b}^T$	dyadic matrix, i.e., a rank one matrix with $(\underline{a}\underline{b}^T)_{mn} = a_m b_n$
$\mathscr{N} = G+B$	partition of the state space into a set G (good) and a set B (bad)
$\mathbf{a}_G$, $\mathbf{a}_{GG}$	submatrix of $\mathbf{a}$ on G×G
$\underline{p}_G$	subvector of $\underline{p}$ on G
$\underline{1}_G$	vector of 1's with cardinality of G
$E[T]$, $\overline{T}$, μ_T	expectation of the random variable T

$\overline{T^K}$	K-th moment of T
σ_T^2	variance of T
$s_T(t)$	probability density function of T
$S_T(t)$	cumulative density function of T, i.e., $S_T(t) = \int_{-\infty}^{t} s_T(x)dx$
$\overline{S}_T(t)$	survival function of T, i.e., $\overline{S}_T(t) = \int_t^{\infty} s_T(x)dx$
$\sigma_T(s), \mathscr{L}\{s_T(t)\}$	the Laplace transform of $s_T(t)$, i.e., $\sigma_T(s) = \int_0^{\infty} e^{-st}s_T(t)dt$
$\pi(s) = \mathscr{L}\{\mathrm{p}(t)\}$	the Laplace transform of the matrix $\mathrm{p}(t)$, i.e., $(\pi(s))_{mn} = \int_0^{\infty} e^{-st}(\mathrm{p}(t))_{mn}dt$, all t
$a(t)*b(t)$	convolution of a(t) and b(t), i.e., $a(t)*b(t) = \int_{-\infty}^{\infty} a(x)b(t-x)dx$ or $a(t)*b(t) = \int_0^t a(x)b(t-x)dx$ when appropriate
$a(t) \sim b(t),\ t \to \infty$	a(t) is asymptotically equal to b(t) as $t \to \infty$, i.e., $a(t)/b(t) \to 1$ as $t \to \infty$

Contents

Page

FOREWORD AND ACKNOWLEDGMENT.............................. v

NOTATION.. vii

CHAPTER 0. INTRODUCTION AND SUMMARY........................ 1

CHAPTER 1. DISCRETE TIME MARKOV CHAINS; REVERSIBILITY IN TIME.................................... 15

§1.00. Introduction........................ 15

§1.0. Notation, Transition Laws........... 15

§1.1. Irreducibility, Aperiodicity, Ergodicity; Stationary Chains....... 16

§1.2. Approach to Ergodicity; Spectral Structure, Perron-Romanovsky-Frobenius Theorem.................. 17

§1.3. Time-Reversible Chains.............. 18

CHAPTER 2. MARKOV CHAINS IN CONTINUOUS TIME; UNIFORMIZATION; REVERSIBILITY........................ 20

§2.00. Introduction........................ 20

§2.0. Notation, Transition Laws; A Review. 20

§2.1. Uniformizable Chains - A Bridge Between Discrete and Continuous Time Chains.............................. 22

§2.2. Advantages and Prevalence of Uniformizable Chains.................. 24

§2.3. Ergodicity for Continuous Time Chains.............................. 25

§2.4. Reversibility for Ergodic Markov Chains in Continuous Time.......... 26

§2.5. Prevalence of Time-Reversibility.... 27

CHAPTER 3. MORE ON TIME-REVERSIBILITY; POTENTIAL COEFFICIENTS; PROCESS MODIFICATION............... 31

§3.00. Introduction........................ 31

§3.1. The Advantages of Time-Reversibility 32

§3.2. The Spectral Representation......... 32

Page

§3.3. Potentials; Spectral Representation 35

§3.4. More General Time-Reversible Chains. 38

§3.5. Process Modifications Preserving Reversibility 38

§3.6. Replacement Processes 41

CHAPTER 4. POTENTIAL THEORY, REPLACEMENT, AND COMPENSATION 43

§4.00. Introduction 43

§4.1. The Green Potential 44

§4.2. The Ergodic Distribution for a Replacement Process 45

§4.3. The Compensation Method 47

§4.4. Notation for the Homogeneous Random Walk 47

§4.5. The Compensation Method Applied to the Homogeneous Random Walk Modified by Boundaries 49

§4.6. Advantages of the Compensation Method. An Illustrative Example..... 51

§4.7. Exploitation of the Structure of the Green Potential for the Homogeneous Random Walk 53

§4.8. Similar Situations 56

CHAPTER 5. PASSAGE TIME DENSITIES IN BIRTH-DEATH PROCESSES; DISTRIBUTION STRUCTURE 57

§5.00. Introduction 57

§5.1. Passage Time Densities for Birth-Death Processes 57

§5.2. Passage Time Moments for a Birth-Death Process 61

§5.3. PF_∞, Complete Monotonicity, Log-Concavity and Log-Convexity 63

§5.4. Complete Monotonicity and Log-Convexity 66

§5.5. Complete Monotonicity in Time-Reversible Processes 67

Page

§5.6. Some Useful Inequalities for the Families CM and PF_∞ 68

§5.7. Log-Concavity and Strong Unimodality for Lattice Distributions 70

§5.8. Preservation of Log-Concavity and Log-Convexity under Tail Summation and Integration 73

§5.9. Relation of CM and PF_∞ to IFR and DFR Classes in Reliability 74

CHAPTER 6. PASSAGE TIMES AND EXIT TIMES FOR MORE GENERAL CHAINS 76

§6.00. Introduction 76

§6.1. Passage Time Densities to a Set of States 77

§6.2. Mean Passage Times to a Set via the Green Potential 81

§6.3. Ruin Probabilities via the Green Potential 84

§6.4. Ergodic Flow Rates in a Chain 86

§6.5. Ergodic Exit Times, Ergodic Sojourn Times, and Quasi-Stationary Exit Times 88

§6.6. The Quasi-Stationary Exit Time. A Limit Theorem 90

§6.7. The Connection Between Exit Times and Sojourn Times. A Renewal Theorem 92

§6.8. A Comparison of the Mean Ergodic Exit Time and Mean Ergodic Sojourn Time for Arbitrary Chains 97

§6.9. Stochastic Ordering of Exit Times of Interest for Time-Reversible Chains. 99

§6.10. Superiority of the Exit Time as System Failure Time; Jitter 102

Page

CHAPTER 7. THE FUNDAMENTAL MATRIX, AND ALLIED TOPICS... 105

§7.00. Introduction........................ 105

§7.1. The Fundamental Matrix for Ergodic Chains.............................. 106

§7.2. The Structure of the Fundamental Matrix for Time-Reversible Chains... 109

§7.3. Mean Failure Times and Ruin Probabilities for Systems with Independent Markov Components and More General Chains...................... 112

§7.4. Covariance and Spectral Density Structure for Time-Reversible Processes........................... 118

§7.5. A Central Limit Theorem............. 121

§7.6. Regeneration Times and Passage Times-Their Relation For Arbitrary Chains. 122

§7.7. Passage to a Set with Two States.... 125

CHAPTER 8. RARITY AND EXPONENTIALITY.................. 130

§8.0. Introduction........................ 130

§8.1. Passage Time Density Structure for Finite Ergodic Chains; the Exponential Approximation................. 131

§8.2. A Limit Theorem for Ergodic Regenerative Processes.................... 133

§8.3. Prototype Behavior: Birth-Death Processes; Strongly Stable Systems.. 137

§8.4. Limiting Behavior of the Ergodic and Quasi-stationary Exit Time Densities and Sojourn Time Densities for Birth-Death Processes.................... 143

§8.5. Limit Behavior of Other Exit Times for More General Chains............ 145

§8.6. Strongly Stable Chains, Jitter; Estimation of the Failure Time Needed for the Exponential Approximation.............................. 150

§8.7. A Measure of Exponentiality in the Completely Monotone Class of Densities........................... 152

Page

§8.8. An Error Bound for Departure from Exponentiality in the Completely Monotone Class.................... 155

§8.9. The Exponential Approximation for Time-Reversible Systems............ 156

§8.10. A Relaxation Time of Interest....... 161

CHAPTER 9. STOCHASTIC MONOTONICITY.................... 164

§9.00. Introduction........................ 164

§9.1. Monotone Markov Matrices and Monotone Chains........................ 164

§9.2. Some Monotone Chains in Discrete Time................................ 168

§9.3. Monotone Chains in Continuous Time.. 171

§9.4. Other Monotone Processes in Continuous Time.......................... 174

REFERENCES.. 176

INDEX.. 181

Chapter 0

Introduction and Summary

§0.0.

This study has three goals. The first is the presentation of working tools needed to quantify the behavior of finite Markov chains in discrete and continuous time when the chain has many degrees of freedom. Ergodic state probabilities, ergodic flow rates, ruin probabilities, passage time and regeneration time distributions and their moments are of typical interest. Applications are oriented largely to reliability theory and inventory theory, but the methods apply as well to other branches of applied probability.

The second goal is the development in detail of topics surrounding reversibility in time, a basic key to tractability. The exploitation of spatial homogeneity and independence underlying certain "modified processes" is also discussed at length.

Finally emphasis is given to rarity and exponentiality, i.e., to the limit theorems concerning these, to measures of exponentiality and error bounds associated with such measures, and to the quantification of departures from exponentiality in failure time distributions for systems modeled by finite chains.

This introductory chapter attempts to provide an overview of the material and ideas covered. The presentation is loose and fragmentary, and should be read lightly initially. Subsequent perusal from time to time may help tie the material together and provide a unity less readily obtainable otherwise. The detailed presentation begins in Chapter 1, and some readers may prefer to begin there directly.

§0.1. Time-Reversibility and Spectral Representation.

Continuous time chains may be discussed in terms of discrete time chains by a uniformizing procedure (§2.1) that simplifies and unifies the theory and enables results for discrete and continuous time to be discussed simultaneously. Thus if $N(t)$ is any finite Markov chain in continuous time governed by transition rates ν_{mn} one may write for $\mathfrak{p}(t) = [p_{mn}(t)] = P[N(t) = n \mid N(0) = m]$

$$\mathfrak{p}(t) = \exp\left[-\nu t(I - \mathfrak{a}_\nu)\right] \tag{0.1.1}$$

where $\nu > \operatorname{Max}_m \sum_n \nu_{mn}$, and $a_{\nu mn} = \nu_{mn}/\nu$ $m \neq n$, $a_{\nu mm} = 1 - \nu^{-1}\sum_n \nu_{mn}$. Hence $N(t) \overset{\text{law}}{=} N^*_{K(t)}$ where N^*_k is governed by $\mathfrak{a}_\nu$, and $K(t)$ is a Poisson process of rate ν independent of N_k.

Time-reversibility (§1.3, §2.4, §2.5) is important for many reasons. A) The only broad class of tractable chains suitable for stochastic models is the time-reversible class. Time-reversibility gives rise to tractability because $\mathfrak{e}_D\mathfrak{p}(t) = [e_m p_{mn}(t)]$ is symmetric and self-adjointness is present with all the simplifications associated therewith. B) Only for the time-reversible class are ergodic probabilities available at once, and only for this class will passage times and exit times of interest have simple structural forms. C) Every birth-death process, "tree" process (§2.5), and vector process $\underline{N}(t)$ with independent time-reversible components is time reversible. The latter processes are appropriate to the description of certain simple random nets with independent links, and complex repairable systems in reliability theory with independent components which are Markov or memoryless by virtue of exponentially distributed failure times and repair times.

For ergodic chains in discrete or continuous time, time-reversibility is equivalent to detailed balance, i.e., to $e_m a_{mn} = e_n a_{nm}$ where $e_m = \lim P[N_k^* = m] = \lim P[N(t) = m]$ is the ergodic probability of state m, or equivalently to $e_m \nu_{mn} = e_n \nu_{nm}$ in continuous time. The notion of time-reversibility, due in large part to Kolmogorov, has meaning for transient Markov chains as well. Here $\pi_D \mathfrak{p}(t) = [\pi_m p_{mn}(t)]$ is symmetric, where the set $\{\pi_m\}$ are the "potential coefficients" available when time-reversibility is present. These topics are discussed in §3.2 and §3.3. Whenever time-reversibility is present one has available the spectral representation (§3.3)

$$p_{mn}(t) = \sqrt{\frac{\pi_n}{\pi_m}} \sum_{j=1}^{N} e^{-\alpha_j t} u_m^{(j)} u_n^{(j)} . \qquad (0.1.2)$$

In matrix form,

$$\mathfrak{p}(t) = \pi_D^{-1/2} \left\{ \sum_{j=1}^{N} e^{-\alpha_j t} J_j^* \right\} \pi_D^{+1/2}$$

where J_j^* is symmetric, and idempotent, with $J_j^* J_k^* = \delta_{jk}$; the α_j are positive except for the ergodic chains where one value of α_j may be zero. This spectral representation is an indispensible tool.

A time-reversible chain $N(t)$ may be modified (§3.5) in a variety of ways without altering the reversibility: states may be made absorbing, transitions may be censored in symmetrical pairs; states may be aggregated; and losses may be introduced. A variety of modified processes may be discussed from this point of view.

§0.2. The Green Potential and the Fundamental Matrix (Chapters 4 and 7).

Let $N(t)$ be a Markov chain with permanent loss to the exterior of the state space of interest. Such chains may be called "lossy" chains, and for such chains, $\sum_n p_{mn}(t)$ is strictly less than one for all m, $t > 0$. Then

$$\mathfrak{g} = \int_0^\infty \mathfrak{p}(t)\, dt \qquad (0.2.1)$$

is finite. This entity which may be called the potential or Green potential for the chain appears intact in a variety of settings. When the chain $N(t)$ is ergodic, so that $\mathfrak{p}(t)$ is stochastic for all t, one may consider in place of $\mathfrak{g}$

$$Z = \int_0^\infty (\mathfrak{p}(t) - \underline{1}\underline{e}^T)\, dt \qquad (0.2.2)$$

where $(\underline{1}\underline{e}^T)_{mn} = e_n$, the ergodic probability for state n. Again Z is finite. In keeping with usage for discrete time chains, Z may be called the fundamental matrix for the chain $N(t)$.

The fundamental matrix plays a key role in

a) the calculation of the mean passage time from a state to a set (§7.6);

b) the asymptotic variance growth rate for the central limit theorem for additive processes with chain dependent growth rate (§7.5);

c) a certain relaxation time which may be defined for an ergodic chain $N(t)$ (§8.10).

Of allied interest are the higher order matrices

$$Z_k = \int_0^\infty t^k(\mathfrak{p}(t) - \underline{1}\underline{e}^T)\, dt. \tag{0.2.3}$$

One then has

$$Z_k = k!\; Z_0^{k+1}. \tag{0.2.4}$$

These matrices enter into the discussion of higher order moments of the random time to a set.

The matrix Z_0 has simple structural properties. For time-reversible chains $\mathfrak{e}_D Z_0$ is positive definite and its null space is of rank one. When one considers a principal submatrix Z_{0G} for some proper subset of the state space, one finds that Z_{0G} is nonsingular.

The green potential $\mathfrak{g}$ is a potential in a traditional sense. If the transient process $N(t)$ is modified by replacing all sample paths going to some specified subset B at a replacement state or on some set of states with a replacement distribution $\underline{r}$, and if the modified process $N^*(t)$ is then ergodic, the ergodic distribution $\underline{e}^{*T}$ is given by

$$\underline{e}^{*T} = \underline{c}^T \mathfrak{g} \tag{0.2.5}$$

where $\underline{c}^T$ is a compensation measure simply related to the dynamics of the replacement process. This point of view used systematically permits a simple real domain treatment of homogeneous processes modified by boundaries.

Such a procedure has a variety of useful benefits.

a) The rank of the problem is often much reduced, since then one deals only with the "boundary" states involved in the modified dynamics of the process.

b) Working with the transition probabilities of the underlying process enables one to exploit the simplification present due to spatial homogeneity or independence of components.

When the underlying process $N(t)$ is ergodic rather than transient, and $N(t)$ is modified by rerouting certain transitions, one finds that the fundamental matrix Z acts as the potential for the process in the same traditional sense. The advantages a) and b) above are again present.

For ergodic processes, the fundamental matrix Z plays the role of the potential, and the ergodic distribution for modified processes may be found with its help. Correspondingly Z appears intact in the mean passage times from a state to a set. For a time-reversible chain $N(t)$, one obtains for $E[T_{0B}]$ the mean time from state 0 to set B,

$$E[T_{0B}] = \frac{1 - \underline{\theta}_0^T \, \theta_0^{-1} \, \underline{1}}{\underline{e}^T \, \theta_0^{-1} \, \underline{1}} \tag{0.2.6}$$

where $\underline{\theta}_0^T = [Z_{0;0m}; m \in B]$, $\theta_0 = [Z_{0;mn}; m;n \in B]$ and $\underline{e}^T$ is the ergodic vector for the subset B.

When $N(t)$ is a stationary ergodic chain, governed by $\mathfrak{p}(t)$, one finds easily that

$$r_f(\tau) = \mathrm{cov}[f(N(t)), f(N(t+\tau)] = \underline{f}^T e_D\{p(t) - \underline{1}\underline{e}^T\}\underline{f}. \tag{0.2.7}$$

Correspondingly for $S(t) = \int_0^t f(N(y))\,dy$ one has

$$\frac{S(t) - \mu t}{\sigma\sqrt{t}} \overset{d}{\to} N(0,1) \tag{0.2.8}$$

where

$$\sigma_f^2 = 2\int_0^\infty r_f(\tau)d\tau = 2\underline{f}^T e_D Z_0 \underline{f} \tag{0.2.9}$$

is the asymptotic variance required for the central limit theorem (0.2.8).

A useful relaxation time may be defined for the process $f(N(t))$ by

$$(T_{REL})_f = \int_0^\infty \frac{r_f(\tau)}{r_f(0)}\,d\tau = \frac{\underline{f}^T e_D Z_0 \underline{f}}{\mathrm{Var}\, f[N(t)]}\,. \tag{0.2.10}$$

A related relaxation time for the chain itself is

$$(T_{REL})_{N(t)} = \underset{f}{\mathrm{Max}}\,(T_{REL})_f.$$

Simple examples and discussion are given in §8.10.

§0.3. Passage Time Densities in Birth-death Processes; Distribution Structure (Chapter 5).

Let $N(t)$ be a well-defined birth-death process on $\mathcal{N} = \{0,1,2,\ldots\}$ governed by $\{\lambda_n,\mu_n\}$ with $\lambda_n > 0$, $n \geq 0$; $\mu_0 = 0$; $\mu_n > 0$, $n \geq 1$. For any such process the passage time density $s_{n,n+1}(\tau)$ from state n to the higher adjacent state is completely monotone in density, as is also $s_{n,n-1}(\tau)$. On the other hand, the passage time density $s_{0n}(\tau)$ from the reflecting state 0 to any state n has the form

$$s_{0n}(\tau) = \theta_{n1}e^{-\theta_{n1}\tau} * \theta_{n2}e^{-\theta_{n2}\tau} * \cdots * \theta_{nn}e^{-\theta_{nn}\tau}, \tag{0.3.1}$$

i.e., is the convolution of exponential densities and belongs to the Polya class PF_∞ [21].

The completely monotone densities are log-convex. The PF_∞ densities are log-concave. These density classes have simple properties of interest for the behavior of more elaborate chains. For any log-convex density $f_X(x)$ one has

$$\left(\frac{\sigma^2}{\mu^2}\right)_X \geq 1. \tag{0.3.2}$$

When the density is log-concave, one has

$$\left(\frac{\sigma^2}{\mu^2}\right)_X \leq 1. \tag{0.3.3}$$

Equality in either case occurs only when $f_X(x)$ is purely exponential. Such structure has wide prevalence in the classical distributions of mathematical statistics. Of similar interest are the "log-concave" and "log-convex" lattice distributions $(p_n)_{-\infty}^{\infty}$, for which

$$p_n^2 \geq p_{n+1}\, p_{n-1}, \quad -\infty < n < \infty \tag{0.3.4}$$

and

$$p_n^2 \leq p_{n+1}\, p_{n-1}, \quad -\infty < n < \infty \tag{0.3.5}$$

and there are no gaps in the intervals of support.

The mean and variance of the passage time densities $s_{mn}(\tau)$ may be calculated explicitly by a self-consistency argument and recursion. These moments, of practical value because of the simplicity of birth-death processes and their wide use as models, also permit quantification of the exponentiality present for passage to high levels in a prototype setting.

§0.4. Passage Times and Exit Times for Ergodic Chains (Chapter 6).

In applications, one is often asked to describe the behavior of a Markov chain $N(t)$ in the following setting. The state space $\mathcal{N}$ consists of two disjoint sets of interest, the good set G and the bad set B. In the study of the reliability of a complex repairable system, for example, G might be the set of states in which the system is working and B might be the failed states. The time at which the set B is first reached from some specified initial distribution $p_n(0)$ on G may be called an exit time from G. A variety of exit times are of interest.

a) One may wish to know the time T_0 to set B from some specific state 0, say the perfect state in the reliability system or the state in which the system is empty in a service system.

b) One may be interested in T_E, the ergodic exit time for which the specified initial distribution is the ergodic distribution truncated to the set G, and renormalized. Here the p.d.f. of T_E is

$$s_E(\tau) = \sum_G e_n s_{nB}(\tau) \Big/ \sum_G e_n \tag{0.4.1}$$

where e_n is the ergodic probability of state n. For the reliability system T_E is the failure time when it is known that the system has always been in existence and is working, and nothing else is known.

c) Of interest is the quasi-stationary exit time T_Q when the initial distribution is the quasi-stationary distribution for which

$$p_n(0) = q_n = \lim_{T\to\infty} P[N(t) = n \mid N(t') \in G,\ t-T \le t' \le t].$$

For the reliability context this would condition the system as working, and having been working into the indefinite past. For this exit time, one has

$$s_Q(\tau) = (E[T_Q])^{-1} \exp\{-\tau/E[T_Q]\}. \qquad (0.4.2)$$

d) A fourth time of special importance is the ergodic sojourn time T_V on G, or post-entry exit time. Here

$$p_n(0) = \frac{i_{Bn}}{\sum_G i_{Bn}} = \frac{\sum_B e_m \nu_{mn}}{\sum_{BG} e_m \nu_{mn}}, \qquad (0.4.3)$$

i.e., one weights the states of G by the relative frequency of entry into G at state n as determined from the ergodic flow rates. The ergodic sojourn time is important because $E[T_V]$ is available when the ergodic probabilities are known. One finds from a simple argument that

$$E[T_V] = \frac{\sum_G e_n}{\sum_{GB} e_m \nu_{mn}} = \frac{P(G)}{i_{BG}}. \qquad (0.4.4)$$

The ergodic exit time density $s_E(\tau)$ and sojourn time density and $s_V(\tau)$ are related by

$$s_E(\tau) = \frac{\int_\tau^\infty s_V(y)\, dy}{E[T_V]} = \frac{\overline{S}_V(\tau)}{E[T_V]}. \qquad (0.4.5)$$

This is also the relationship between any lifetime density in renewal theory and the density of the time until the next renewal at stationarity. In this sense, (0.4.5) may be regarded as a generalized renewal theorem for chains in continuous time and more general ergodic processes for which $s_E(\tau)$

and $\bar{S}_V(\tau)$ are meaningful.

It follows from (0.4.5) that for any chain $N(t)$ in continuous time one has

$$\bar{T}_E \geq \frac{1}{2} \bar{T}_V. \qquad (0.4.6)$$

When the underlying process $N(t)$ is reversible in time, more can be said. An argument based on the spectral representation (0.1.2) shows that T_Q, T_E, and T_V are completely monotone in density. It then may be shown that one has the stochastic order relation

$$T_Q \succ T_E \succ T_V \qquad (0.4.7)$$

where $X \succ Y$ means that $F_X(x) \leq F_Y(x)$ for all x.

$$E[T_Q] \geq E[T_E] \geq E[T_V] \qquad (0.4.8)$$

and

$$P[T_E > x] \leq \exp\{-x/E[T_Q]\}. \qquad (0.4.9)$$

This bound is of practical interest because it involves the single parameter $E[T_Q]$, which may be approximated in many ways. These equations and bounds are of interest in reliability theory for the study of repairable systems with independent memoryless components whose time-reversibility has been noted.

§0.5. Exponential Limit Theorems for Exit Times of Interest. Measures of Exponentiality (Chapter 8).

The intuitive basis for exponentiality in the presence of rarity may be understood in the following setting. Suppose, for example, one has an ergodic Markov chain $N(t)$ on a denumerably infinite state space $\mathcal{N}$. Consider a sequence of nested partitions $\mathscr{N} = G_K + B_K$; $G_K \subset G_{K+1}$; $K = 1,2,\ldots$ Let $n = 0 \in G_1$, and let $E[T_{0B_K}] \to \infty$, as $K \to \infty$, where T_{0B_K} is the passage time from 0 to B_K. The limit theorem [26] states that $T_{0B_K} / E[T_{0B_K}] \xrightarrow{d} Y$, where $F_Y(x) = 1 - e^{-x}$. The onset of exponentiality is associated with the regenerative character of the ergodic chain. For $N(t)$, the mean return time to state 0 is finite. As $K \to \infty$ and B_K becomes more remote, the probability of regenerating by returning to state 0 before reaching B_K goes to one. Consequently, there will be more and more regenerations before reaching B_K as $K \to \infty$. Each such regeneration may be regarded as completing a trial, and one has in effect a probability per unit time of reaching B_K for K large. The last leg of the journey, the direct passage to B_K from state 0 without returning to state 0 has a mean time which is asymptotically negligible compared to the time until the last departure from state 0. The latter is the crux of the proof.

For any pair of states m and n in G_K one has the triangle inequality

$$E[T_{mB_K}] \leq E[T_{mn}] + E[T_{nB_K}] \tag{0.5.1}$$

and the same inequality with m, n reversed. It follows that when m, n are held fixed

$$E[T_{mB_K}] \sim E[T_{nB_K}], \quad K \to \infty. \tag{0.5.2}$$

Indeed, the triangle inequality has a stronger form $T_{mB_K} \prec T_{mn} + T_{nB_K}$, as shown jointly with D. R. Smith. These tools then imply that $E[T_{EB_K}] \sim E[T_{mB_K}]$ for any fixed m as $K \to \infty$ where T_{EB_K} is the ergodic exit time from B_K. Moreover, one has, as $K \to \infty$, the limit behavior

$$\frac{T_{EB_K}}{E[T_{EB_K}]} \xrightarrow{d} Y; \quad F_Y(x) = 1 - e^{-x}. \tag{0.5.3}$$

One can also show that for time-reversible chains $E[T_{QB_K}] \sim E[T_{EB_K}]$ and that $T_{QB_K}/E[T_{QB_K}] \to Y$; as $K \to \infty$. In this context, the time from any fixed state, the ergodic exit time and the quasi-stationary exit time all have the same expectation asymptotically and have the same limit behavior.

In contrast to the central limit theorem for sums of i.i.d. random variables, convergence to exponentiality with increasing rarity of the set visited is usually rapid. To quantify the departure from exponentiality, a measure of exponentiality [§8.7] and associated error bound [§8.8] are available when time-reversibility is present. The measure of exponentiality in the completely monotone density class is for density $f_X(x)$,

$$\zeta_X = \left(\frac{\sigma^2}{\mu^2}\right)_X - 1. \tag{0.5.4}$$

In this class one then has the error bound, when $EX = 1$,

$$|P[X > x] - e^{-x}| \le K\,(\sigma_X^2 - 1)^{1/4}$$

where K is some positive constant $\le 8\pi^{-1}(4.5)^{1/4}$. In Chapter 8, these limit theorems and error bounds are discussed and the validity of the exponential approximation for ergodic systems of independent memoryless components is explored.

0.6. Stochastically Monotone Chains (Chapter 9).

Many of the Markov chains in continuous time on an ordered state space have the property that the state vector $\underline{p}(t)$ with components $p_{0n}(t) = P[N(t) = n \mid N(0) = 0]$ grows stochastically in time, i.e.,

$$\underline{p}(t_2) \succ \underline{p}(t_1), \quad t_2 \geq t_1. \qquad (0.6.1)$$

In particular, all birth-death processes have this behavior and all spatially homogeneous processes modified by retaining (impenetrable) or absorbing boundaries have this behavior. The stochastic monotonicity of the state vector (0.6.1) has its origin in a corresponding more basic property of the operators generating transitions [10,20]. Such "monotone operators" have a dual characterization,

a) They preserve stochastic order under post-multiplication, i.e., $\underline{p}_1 \succ \underline{p}_2 \Rightarrow \underline{p}_1\mathfrak{a} \succ \underline{p}_2\mathfrak{a}$, when $\mathfrak{a}$ is monotone.

b) They preserve component monotonicity under pre-multiplication, i.e., $\underline{f} = [f_n] : f_n\uparrow \Rightarrow (\mathfrak{a}f)_n\uparrow_n$.

This simple property is important because of the many insights it generates in reliability theory and other applied probability settings.

Chapter 1

Discrete Time Markov Chains; Reversibility in Time

§1.00.

The first chapter is devoted to a quick review of the elements [14,15] of Markov chains in discrete time. The topic of reversibility in time, of basic importance for what follows is discussed at greater length.

§1.0. Notation, Transition Laws.

Let $(N_k)_0^\infty$ be a sequence of random variables (r.v.'s) on the *state space* $\mathscr{N} = \{0,1,2,...\}$, finite or infinite.

Let the process $(N_k: k \geq 0)$ be: a) *Markov*; b) *temporally homogeneous*, i.e. the laws governing transitions are fixed in time. Then $(N_k : k \geq 0)$ is governed by a single step transition matrix $\mathbf{a} = [a_{mn}]$, with $a_{mn} \geq 0$, $\Sigma_n a_{mn} = 1$, and $a_{mn} = P[N_{k+1} = n | N_k = m]$, independent of k. The *k-step transition probability matrix* is denoted by $\mathbf{a}^{(k)} = [a_{mn}^{(k)}]$, with $a_{mn}^{(k)} = P\{N_{r+k} = n | N_r = m\}$. From a) and b) one has $\mathbf{a}^{(k)} = \mathbf{a}^k$, the k-th power of $\mathbf{a}$.

The distribution of N_k is given by the *probability vector* $\underline{p}_k^T = (p_0(k), p_1(k),...)$, with $p_m(k) = P[N_k = m]$.

Of particular interest is the initial distribution $\underline{p}_0^T$. We have $\underline{p}_k^T = \underline{p}_0^T \mathbf{a}^k$ or component-wise $p_n(k) = \Sigma_m p_m(0) a_{mn}^{(k)}$.

Self-consistency requires the *Chapman - Kolmogorov* equations; $a_{mn}^{(k_1+k_2)} = \Sigma_r a_{mr}^{(k_1)} a_{rn}^{(k_2)}$. In matrix notation, this is simply $\mathbf{a}^{(k_1+k_2)} = \mathbf{a}^{(k_1)} \mathbf{a}^{(k_2)}$, a direct consequence of $\mathbf{a}^{(k)} = \mathbf{a}^k$.

§1.1. Irreducibility, Aperiodicity, Ergodicity; Stationary Chains.

Definitions: Let N_k be a discrete time Markov chain.

N_k is *irreducible* if $a_{mn}^{(k)} > 0$ for some k for all m,n. (All states are said to "communicate", i.e. every state can be reached from every other state in a finite number of steps.)

N_k is *aperiodic* if G.C.D. $\{k | a_{mn}^{(k)} > 0\} = 1$ all m,n. (G.C.D. = greatest common divisor). Irreducibility and aperiodicity are contained in the block structure of matrix $\mathbf{a}$, i.e. in its configuration of zeros.

N_k is *ergodic* if $\lim_{k\to\infty} a_{mn}^{(k)} = e_n > 0$ all m. (in this case $e_n = \lim_{k\to\infty} P[N_k = n]$; $\underline{p}_0^T$ does not affect the limiting distribution). N_k ergodic implies $\underline{e}^T \mathbf{a} = \underline{e}^T$.

The *first passage time from* m *to* n τ_{mn} is defined by $\tau_{mn} = \min_{k>1} \{k: N_k = n | N_0 = m\}$. Also $\tau_{mn} = \min_{k\geq 1}\{k: N_{k+r} = n | N_r = m\}$, from temporal homogeneity.

State m is *recurrent* if $P[\tau_{mm} < \infty] = 1$, i.e. τ_{mm} is a proper r.v. Given recurrence, m is *positive recurrent* if $E\tau_{mm} < \infty$, *null recurrent* if $E\tau_{mm} = \infty$. m is *transient* if $P[\tau_{mm} < \infty] < 1$.

These state properties are class properties: if all states are communicating and there are no transient states, then the process itself can be said to be *positive recurrent* or *null recurrent*.

Of particular importance is the following theorem.

Theorem: N_k is ergodic iff N_k is irreducible, aperiodic and positive recurrent. In particular, if $\mathcal{N}$ is finite, N_k is ergodic iff N_k is irreducible and aperiodic.

N_k is said to be *stationary* if N_k is ergodic and $\underline{p}_0^T = \underline{e}^T$. Stationarity implies $\underline{p}_k^T = \underline{p}_0^T = \underline{e}^T$, all k.

§1.2. Approach to Ergodicity; Spectral Structure; The Perron-Romanovsky-Frobenius Theorem.

Assume N_k ergodic, i.e., $\exists \underline{e}^T: \underline{e}^T a = \underline{e}^T$, and $\lim_{k\to\infty} a^{(k)} = \underline{1}\,\underline{e}^T \overset{\text{def}}{=} J$. Here J is a dyadic matrix of rank 1 with $(\underline{1}\,\underline{e}^T)_{mn} = e_n$. $\underline{e}^T$ is a left eigenvector of a, with eigenvalue 1; the corresponding right eigenvector is $\underline{1}$: $a\,\underline{1} = \underline{1}$. One has, immediately,

$$J\,a = J, \quad a\,J = J, \quad J^2 = J \quad \text{(idempotent)}. \tag{1.2.1}$$

If we introduce Δ by $a = J+\Delta$ then $\Delta\underline{1} = \underline{0}$, and from (1.2.1)

$$\Delta J = J\Delta = 0. \tag{1.2.2}$$

It follows from (1.2.1), (1.2.2) that

$$a^k = (J+\Delta)^k = J+\Delta^k, \tag{1.2.3}$$

so that $\lim_{k\to\infty} \Delta^k = 0$ (component-wise). These results hold whether $\mathcal{N}$ is finite or infinite. Fubini's theorem may be invoked where needed.

Now suppose $\mathscr{N}$ to be finite so that a is an $N \times N$ matrix. Let the set of eigenvalues of a be $\{\lambda_j(a)\}_{j=1}^{N}$. For this finite case, the *Perron-Romanovsky-Frobenius theorem* (see, e.g., [12]) states that $\max_j\{|\lambda_j(\Delta)|\} = r < 1$. One then has $\max_j\{|\lambda_j(\Delta^k)|\} = r^k \to 0$ as $k \to \infty$.

§ 1.3. Time-Reversible Chains.

Let a be ergodic, (by this we mean that a is the single-step transition matrix of an ergodic process N_k), and consider $a_R \overset{\text{def}}{=\!=} e_D^{-1} a^T e_D$, where $e_D = \text{diag}(e_0, e_1, \ldots)$, the diagonal matrix with the ergodic probabilities as its non-zero elements. Then $a_R\underline{1} = e_D^{-1} a^T e_D\underline{1} = e_D^{-1} a^T \underline{e} = e_D^{-1}\underline{e} = \underline{1}$, so that $a_R\underline{1} = \underline{1}$. Further $a_R \geq 0$, so that a_R is a Markov (i.e., stochastic) matrix. The matrix a_R will be called the *dual* of a. Clearly, every ergodic a has a unique dual. The reader will have no difficulty verifying that a_R is ergodic when a is, and has the same ergodic vector, i.e., $\underline{e}_R^T = \underline{e}^T$. Consequently, $(a_R)_R = a$, and a_R is then a dual in the customary mathematical sense. (The dual of an ergodic chain may be defined correspondingly.)

<u>Definition 1.3A</u>. An ergodic Markov chain is said to be reversible in time if $a_R = a$, i.e., if its governing matrix is its own dual.

We see that for reversible chains $a_R = a \longleftrightarrow e_D^{-1} a^T e_D = a \longleftrightarrow a^T e_D = e_D a \longleftrightarrow (e_D a)^T = e_D a$. Hence

<u>Proposition 1.3B</u>. An ergodic chain is reversible in time if and only if $e_D a$ is symmetric.

The symmetry of $e_D a$ corresponds to *"detailed balance"* for the chain, since one then has $e_m a_{mn} = e_n a_{nm}$. The term on the left is just the mean number of transitions per unit time from m to n for the stationary process, and the term on the right is that for transitions from n to m.

Pre- and post-multiplication of a symmetric matrix by another symmetric matrix preserves symmetry. Therefore

$$e_D a \text{ symmetric} \Longleftrightarrow e_D^{1/2} a e_D^{-1/2} \text{ symmetric}, \qquad (1.3.1)$$

where

$$e_D^{1/2} = \text{diag}(e_0^{1/2}, e_1^{1/2}, \ldots).$$

Since $(e_D^{1/2} a\, e_D^{-1/2})^k = e_D^{1/2}\, a^k\, e_D^{-1/2}$ is symmetric, it follows that $e_D\, a^k$ is symmetric for all k and hence

$$e_m\, a_{mn}^{(k)} = e_n\, a_{nm}^{(k)}, \qquad (1.3.2)$$

for all m,n, all k; for any ergodic time-reversible chain.

The name "time-reversible chain" has its origins in equation (1.3.2), which implies that when N_k is stationary we have

$$P[N_r = m,\ N_{r+k} = n] = P[N_r = n,\ N_{r+k} = m], \qquad (1.3.3)$$

i.e., equality for the joint probabilities for every time separation k. The reader will verify alternately that

$$P[N_{r+k} = n \mid N_r = m] = P[N_r = n \mid N_{r+k} = m]. \qquad (1.3.4)$$

Chapter 2

Markov Chains in Continuous Time; Uniformization; Reversibility

§2.00.

A Markov chain in continuous time may be treated as a Markov chain in discrete time with random Poisson transition epochs, or as a limit of such discrete time chains. The uniformization procedure establishing a theoretical bridge between discrete and continuous time is simple and has a variety of benefits, computational and theoretical. In particular ergodicity and time-reversibility for Markov chains may be discussed in this way. The prevalence and importance of time-reversible chains is examined.

§2.0. Notation, Transition Laws; A Review.

Consider a temporally homogeneous Markov chain $N(t)$ in continuous time, with state space $\mathcal{N} = \{0,1,2,\ldots,N\}$ where N may be finite or infinite, as in §1.0. In place of the single-step transition probabilities for discrete time, the process is governed by a set of hazard rates $\{\nu_{mn}\}$, in the following way: Let the *transition probabilities at time* t be denoted by

$$p_{mn}(t) = P[N(t) = n \mid N(0) = m] \qquad (2.0.1)$$
$$= P[N(t+s) = n \mid N(s) = m],$$

so that $p_{mn}(0) = \delta_{mn}$, where $\delta_{mn} = 0$, $m \neq n$, $\delta_{mm} = 1$. Then for small t one has, by definition

$$p_{mn}(t) = \nu_{mn} t + o(t) \quad \text{as} \quad t \downarrow 0 \text{ for } m \neq n, \qquad (2.0.2a)$$

$$p_{mm}(t) = 1 - \nu_m t + o(t) \quad t \downarrow 0. \qquad (2.0.2b)$$

Here $\nu_m = \Sigma_n \nu_{mn}$ $(\nu_{mm} = 0)$, $o(t)$ denotes a function $f(t)$ with the property $\lim_{t \to 0} \frac{f(t)}{t} = 0$. The hazard rate ν_{mn} is the proability per unit time of going to state n, when the process is in state m. ν_m is the net hazard rate for transition out of state m, i.e., the dwell time of the process in state m is exponentially distributed with density $\nu_m \exp(-\nu_m \tau)$. The transition from state m will carry the process to state n with probability ν_{mn}/ν_m.

At time t, the process will be described by the state probability vector $\underline{p}^T(t) = (p_0(t), p_1(t), \ldots)$ where $p_n(t) = P[N(t) = n]$. Then for initial state probability vector $\underline{p}_0^T = \underline{p}^T(0)$, one has $\underline{p}^T(t) = \underline{p}_0^T \, \mathbf{p}(t)$ where $\mathbf{p}(t) = [p_{mn}(t)]$. From $\underline{p}^T(t+\delta) - \underline{p}^T(t) = \underline{p}^T(t)[\mathbf{p}(\delta) - I]$, and (2.0.2a,b) one obtains

$$\frac{d}{dt} p_n(t) = -\nu_n p_n(t) + \Sigma_m p_m(t) \nu_{mn}. \qquad (2.0.3)$$

These equations are called "forward" equations (cf. Feller II). Our treatment of chains in continuous time will be constructive and will avoid the need for concern with forward equations, "backward" equations (ibid) and behavior at infinity. It will mirror the discussion of discrete time chains, where such concerns are totally unnecessary.

§2.1. Uniformizable Chains - A Bridge Between Discrete Time Chains and Continuous Time Chains.

When the set of hazard rates $\{\nu_m\}$ is bounded, i.e., when $\nu_m \leq \nu < \infty$ for all m the process is said to be "uniformizable". The process with which we will be dealing will either be uniformizable or will be the limit in law of a sequence of such uniformizable processes. If the chain is finite, uniformizability is immediate. Also, for random walks in continuous time on the lattice of integers where $\nu_{mn} = f(n-m)$, and $\mathcal{N} = \{\ldots,-2,-1,0,1,2,\ldots\}$ one has $\nu_m = \Sigma_{-\infty}^{\infty} f_g$ which is independent of m. Clearly its constant value must be finite for the process to be of interest. Other examples will be given in §2.2.

Consider a uniformizable chain, for which $\nu_0 = \sup \nu_m = \sup \Sigma_n \nu_{mn} < \infty$. It will be seen that with probability one, only a finite number of transitions will occur in $[0,t]$, for any $t > 0$. In fact, by adding dummy transitions from states to themselves, a Poisson process can be constructed which governs the epochs at which transitions take place. With the help of this Poisson process the continuous time process is reduced to a discrete time process, where the number of transition epochs is randomized.

To see this, choose *any* $\nu \geq \nu_0$. The forward equations

$$\frac{d}{dt} p_n(t) = -\nu_n p_n(t) + \Sigma_m p_m(t) \nu_{mn}, \qquad (2.1.1)$$

can be rewritten as

$$\begin{aligned} \frac{d}{dt} p_n(t) &= -\nu p_n(t) + \Sigma_m p_m(t) \nu_{mn} + p_n(t) (\nu-\nu_n) \\ &= -\nu\, p_n(t) + \nu\, \Sigma_m p_m(t) a_{mn}, \end{aligned} \qquad (2.1.2)$$

where

$$\begin{cases} a_{mn} = \nu_{mn}/\nu & \text{for } m \neq n \\ a_{mm} = 1 - \nu_m/\nu. \end{cases} \tag{2.1.3}$$

Then $a_{mn} \geq 0$, and $\Sigma_n a_{mn} = 1$, so that a_ν is stochastic. In vector notation (2.1.2) can be written as

$$\frac{d}{dt} \underline{p}^T(t) = -\nu \underline{p}^T(t)[I-a_\nu]. \tag{2.1.4}$$

This has the unique solution $\underline{p}^T(t) = \underline{p}^T(0) e^{-\nu t[I-a_\nu]}$. Consequently, the transition matrix is given by

$$p(t) = e^{-\nu t[I-a_\nu]} = \Sigma_{k=0}^{\infty} e^{-\nu t} \frac{(\nu t)^k}{k!} a_\nu^k. \tag{2.1.5}$$

Note the freedom we have in choosing ν. Only $\nu > \nu_0$ is required. This freedom was made possible by permitting transitions to the same state. Probabilistically, the above construction is equivalent to the following: with the continuous time process $N(t)$, governed by $[\underline{p}_0^T, (\nu_{mn})]$, consider an associated discrete time process N_k^*, governed by $[\underline{p}_0^T, a_\nu]$, and let $K_\nu(t)$ be a Poisson process with parameter ν. Take

$$N(t) = N^*_{K_\nu(t)}, \tag{2.1.6}$$

with N_k^*, $K_\nu(t)$ independent processes. Then $N(t)$ and the original process $N(t)$ will have their state probabilities governed by the same equation (2.0.3) and will be identical in law.

The Poisson process $K_\nu(t)$ for the number of transition epochs in $[0,t]$ has an honest distribution, the Poisson distribution. Consequently, for the set of process sample paths this number will be finite with probability one. As a result, the boundary at infinity, if any, is never reached, and boundary conditions at infinity are not relevant for uniformizable processes. (The importance of such conditions when infinity can be reached in finite time is the raison d'etre for Feller's boundary theory in the classification of birth-death and diffusion processes [7] [24] [50].)

§2.2. Advantages and Prevalance of Uniformizable Chains.

Equations (2.1.1) form a system of linear first-order differential equations. If $\nu < \infty$, these equations have the explicit solution, (cf. 2.1.5) given by

$$p_{mn}(t) = \sum_{k=0}^{\infty} e^{-\nu t} \frac{(\nu t)^k}{k!} a_{mn}^{(k)}, \quad m,n \in \mathcal{N}. \qquad (2.2.1)$$

Here the $a_{mn}^{(k)}$ can be easily machine computed if $\mathcal{N}$ is finite. For infinite $\mathcal{N}$, the necessary iteration and approximation of the $a_{mn}^{(k)}$ does not pose serious difficulties, although convergence rate difficulties may be present.

The decomposition (2.1.5) has the added advantage of permitting study of the properties of $\mathfrak{p}(t)$ in terms of those of the matrix $\mathfrak{a}_\nu$, a mathematically simpler entity. A good example will be seen later when the stochastic monotonicity of birth-death processes is discussed. Ergodicity and time-reversibility will soon be studied in this way.

The following examples show the prevalence of uniformizable chains.

a) *Finite Markov chains* are trivially uniformizable.

b) *Random Walks.* $\mathcal{N}$ consists of a multidimensional lattice, and $\nu_{\underline{m}\underline{n}} = f(\underline{n}-\underline{m})$, so that the process is *spatially homogeneous*, i.e., the increment distribution is independent of the state of the process. Here $\nu_{\underline{m}} = \Sigma_{\underline{n}} \nu_{\underline{mn}}$ is independent of $\underline{m}$, hence the process is uniformizable.

c) *Birth-Death processes.* When $\mathscr{N} = \{0,1,2,\ldots\}$, $\nu_{m\,m+1} = \lambda_m > 0$, $\nu_{m+1\,m} = \mu_{m+1} > 0$ for $m = 0,1,\ldots$, all other $\nu_{nm} = 0$, we have a birth-death process with a reflecting boundary at $n = 0$. For such a process, $\nu_m = \lambda_m + \mu_m$, and the process is uniformizable iff both (λ_m) and (μ_m) are bounded sequences. The *homogeneous birth-death process,* where $\lambda_m = \lambda$, $m \geq 0$, and $\mu_m = \mu$, $m \geq 1$, also known as the *Poisson queue,* is an example.

§2.3. Ergodicity for Continuous Time Chains.

The basic definition of ergodicity for Markov chains in continuous time is the following.

Definition 2.3A. A continuous time process $N(t)$ is *ergodic* if $\lim_{t\to\infty} p_{mn}(t) = e_n > 0$, all $m,n \in \mathscr{N}$. In this case we also say that $\mathfrak{p}(t)$ is ergodic (cf. §1.1).

When a chain is uniformizable, the question of ergodicity may be reduced to a simpler question in discrete time, as in the next theorem.

Theorem 2.3B. If $\mathfrak{a}_\nu$ is ergodic, then $\mathfrak{p}(t) = e^{-\nu t[I-\mathfrak{a}_\nu]}$ is ergodic, and $\mathfrak{a}_\nu$ and $p(t)$ have the same ergodic probabilities. The proof is left to the reader.

When the chain is finite, it is well known that ergodicity is established when the chain is shown to be irreducible [14]. The concept of periodicity and aperiodicity of interest for discrete time chains is no longer present, and the positive recurrence of all states is assured by the finiteness of the state space. Explicitly:

Theorem 2.3C. Every irreducible finite continuous time Markov chain is ergodic.

§2.4. Reversibility for Ergodic Markov Chains in Continuous Time.

As for the discussion of reversibility for discrete time processes in §1.3, we will consider reversibility in the context of ergodic processes. Reversibility is meaningful and useful for transient processes as well. Such reversibility for transient chains will be treated later.

Let $N(t)$ be an ergodic process.

Definition 2.4A. $N(t)$ is *reversible in time* if $\mathfrak{e}_D\, \mathfrak{p}(t)$ is symmetric for all $t > 0$.

Consider a stationary reversible process, i.e., specify $\underline{p}_0^T = \underline{e}^T$. Then $e_m p_{mn}(t) = P\{N(s) = m,\ N(t+s) = n\} = P\{N(s) = n,\ N(t+s) = m\} = e_n p_{nm}(t)$. This equality is generalizable to any finite time skeleton: the stationary process appears the same in all laws, seen forward or backward in time.

The discussion of reversibility in continuous time may be simplified with the help of uniformization, and then related to the theory developed in Chapter 1 for discrete time processes. Specifically

Proposition 2.4B. If N_k, governed by $\mathfrak{a}_\nu$ is reversible in time, then $N(t)$, governed by $[\nu, \mathfrak{a}_\nu]$ is reversible in time.

Proof: $\mathfrak{e}_D\, \mathfrak{p}(t) = \Sigma_{k=0}^{\infty} e^{-\nu t} \frac{(\nu t)^k}{k!} (\mathfrak{e}_D\, \mathfrak{a}_\nu^k)$ is symmetric, a direct consequence of symmetry of $\mathfrak{e}_D\, \mathfrak{a}_\nu^k$ for all k. (cf. (1.3.2))

Proposition 2.4C. Suppose $e_D\, p(t)$ is symmetric and $\nu_m < \nu$. Then $e_D\, a_\nu$ is symmetric.

Proof: The symmetry of $e_D\, e^{-\nu t[I-a_\nu]}$ implies the symmetry of $e_D\, e^{\nu t a_\nu}$ and that of $e_D[e^{\nu t a_\nu}-I]/\nu t$. Using $(e^{\theta a}-I)/\theta \to a$ for $\theta \downarrow 0$, and the fact that symmetry is preserved under limiting operations, we see that $e_D\, a_\nu$ is symmetric.

Theorem 2.4D. Let $N(t)$ be ergodic, and uniformizable. Then $N(t)$ is reversible in time iff

$$e_m \nu_{mn} = e_n \nu_{nm}. \tag{2.4.1}$$

This follows from (2.1.3), 2.4B, and 2.4C. The term $e_m \nu_{mn}$ is the stationary flow rate from m to n. A more precise discussion will be given subsequently. In (2.4.1) we recognize the condition of *detailed balance*. As a direct result of conditions (2.4.1), any chain with a transition going *one way* cannot be reversible in time. As an example, the process with three states and hazard rates shown cannot be reversible because $\nu_{32} > 0$ but $\nu_{23} = 0$. Nevertheless, as we will see

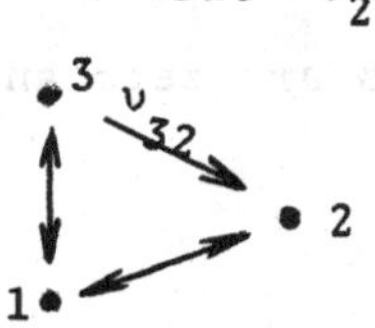

$$\nu_{12}, \nu_{21}, \nu_{13}, \nu_{31} > 0;$$
$$\nu_{32} > 0;\ \nu_{23} = 0.$$

in the next section and §3.3, a considerable class of processes are reversible in time.

§2.5. Prevalence of Time-Reversibility.

a. Ergodic birth-death processes.

Consider the process (c) of §2.2, and suppose the birth-death process is ergodic. For the state space as indicated below an imaginary line may be drawn between any adja-

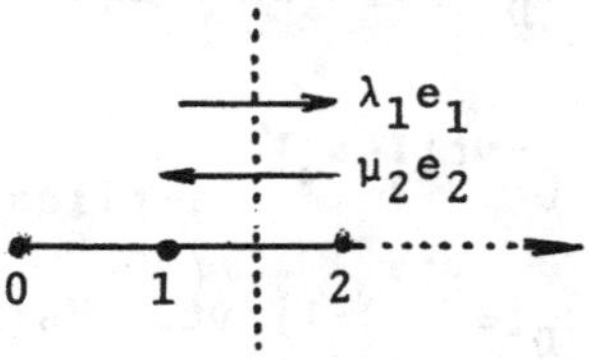

cent states partitioning the state space. When the process is stationary, the net flow from $\{0,1\}$ to $\{2,3,4,\ldots\}$ for example, must be zero since the total probability mass in each subset is constant. Clearly, the system has detailed balance, and every ergodic birth-death process is reversible in time.

A formal argument for the birth-death process with the same content is the following. The forward equations (2.0.3) take the form for $m \geq 1$

$$\frac{d}{dt} p_m(t) = -(\lambda_m+\mu_m)\, p_m(t) + \lambda_{m-1}\, p_{m-1}(t) + \mu_{m+1}\, p_{m+1}(t),$$

with $(d/dt)\, p_0(t) = -\lambda_0 p_0(t) + \mu_1 p_1(t)$. Summation of these equations for $m = 0,1,2,\ldots\ M$ gives

$$\frac{d}{dt}\left(\sum_0^M p_m(t)\right) = -\lambda_M\, p_M(t) + \mu_{M+1}\, p_{M+1}(t)$$

For the stationary process both sides are zero and this implies detailed balance.

b. <u>Ergodic tree processes</u>.

A process may be said to be a *tree process* if its state space is a tree in the sense of graph theory [6], i.e., if i) all transitions are to adjacent states, ii) adjacent states communicate, iii) there are no "loops" possible in the flow. More precisely, if $(N_1,N_2,N_3,\ldots,N_k)$ is a path for the chain with $\nu_{n_i n_{i+1}} > 0$, every such path from N_k back to N_1 must pass through $N_2, N_3, \ldots N_{k-1}$ at least once. Alternately, there is a unique minimal path connecting every pair of states. A typical tree process is shown in A below.

A B

Birth-death processes are special types of tree process. A second particularly simple type of tree process of applied interest is the star process shown in B, where all states communicate through a unique focal state.

The reversibility in time of ergodic tree processes is demonstrated through a detailed balance argument in the same way as for birth-death processes, since every line separating two adjacent states partitions the state space into two disjoint sets.

c. Multivariate processes with independent Markov Components.

Let $N(t) = [N_1(t), N_2(t), \ldots, N_K(t)]$, where $N_i(t)$ are independent Markov chains, time-reversible, and ergodic. The detailed balance property is inherited from the components as follows (for convenience, let $K = 2$): $\nu_{\underline{m}\underline{n}} = \nu_{(m_1,m_2)(n_1,n_2)} = 0$, unless $m_1 = n_1$ or $m_2 = n_2$, since the process with probability one has no transitions in which more than one component changes. Since $e_{\underline{m}}$ factors, i.e., $e_{\underline{m}} = e_{m_1} e_{m_2}$; $e_{\underline{m}} \nu_{\underline{m}\underline{n}} = e_{\underline{n}} \nu_{\underline{n}\underline{m}}$ follows from detailed balance in each of the components.

A special case of interest to reliability theory is that of a system of k independent components, each of which has a constant hazard rate for failure, and a constant hazard rate for repair, so that their failure times and repair times are exponentially distributed. The states of the system are $(n_1, n_2, \ldots n_K)$, where $n_j = 0$, or 1, for the failed state and working state of component j, respectively. By the reasoning above the system state $\underline{N}(t)$ is a finite, ergodic, time-reversible chain. Details are given in [32]. A similar related context in the setting of inventory theory is described in [35].

It will be seen in the next chapter that time-reversible chains may be altered in a variety of ways without destroying the reversibility. For example, the process $[N_1(t), N_2(t)]$ where $N_1(t)$ and $N_2(t)$ are birth-death processes, may be restricted to the states $\mathcal{N}^* = \underline{n} = (n_1, n_2);$ $n_1 \geq 0, n_2 \geq 0, n_1+n_2 < L\}$ by setting all hazard rates leading out of $\mathcal{N}^*$ to the value zero, i.e., by censoring such transitions. The resulting process is still time-reversible.

d. Other settings.

An excellent discussion of time-reversibility and its implications for applied probability models has been given by J.F.C. Kingman [48]. Other useful applications have been described by F. P. Kelly [44], [45].

Chapter 3

More on Time-Reversibility; Potential Coefficients; Process Modification

§3.00.

The significance of the presence of time-reversibility for applied models is emphasized; namely their tractability, the accessibility of their ergodic state probabilities, and the simplicity of their time dependent behavior. The spectral representation of the transition probability matrix $\mathfrak{p}(t)$ is established, and its structure explored. For time-reversible processes, a key property of path independence is present which permits extension of the notion of time-reversibility to transient chains and "lossy" chains. Time-reversible processes may be modified in a variety of ways without destroying the reversibility. The modified process is often of interest in its own right or is of primary interest. The last section introduces replacement processes, of special interest.

In this and subsequent chapters we will be working freely back and forth between discrete and continuous time chains, with the link of uniformization established in Chapter 2 fully in mind. Specifically, when a section appears to be talking about chains in discrete time, it is addressing itself simultaneously to the continuous time counterparts. The latter are usually of greater interest in applications.

§3.1. The Advantages of Time-Reversibility.

The time-reversible processes are a broad and structurally simple class of processes which are *tractable*, i.e., which lend themselves to the kind of theoretical and numerical analysis required in applied problems where quantification is essential. Indeed these processes are, to the author's knowledge, the only such broad class of tractable processes.

They are tractable in that *their ergodic distributions, as we will see, are simply evaluable*, and in that a *spectral representation of the time dependent behavior is present for which the eigenvalues and eigenvectors are real.* This reality has important structural consequences for the distributions describing transient behavior of interest.

Sometimes, for special reasons, non-reversible processes may be assayed conveniently. Consider, for example, the ergodic distribution of queues in tandem. Here a Poisson stream of customers passes through a sequence of servers with independent exponentially distributed service times at each. The ergodic distribution is available for a special subtle reason [58]. But the transient behavior is unavailable analytically, and unavailable numerically for large systems.

§3.2. The Spectral Representation.

It will be assumed unless otherwise stated that we have a finite state space and irreducibility. Much of the structural insight available is retained in the denumerably infinite case.

First, let N_k be a time-reversible ergodic discrete time chain. Then, as shown in §1.3, $e_D a$, and $e_D a^k$ are real symmetric. Hence $e_D^{1/2}\, a\, e_D^{-1/2}$ is symmetric and therefore has a complete set of orthonormal eigenvectors $(\underline{u}_j)_1^N$ associated with *real* eigenvalues $(\lambda_j)_1^N$. Since the eigenvalues of $e_D^{1/2}\, a\, e_D^{-1/2}$ are those of a, from the Perron-Romanovski-Frobenius theorem, [12] we infer that $-1 < \lambda_j < 1$, for $j = 2,3,..N$ and $\lambda_1 = 1$. Hence we may write

$$e_D^{1/2}\, a\, e_D^{-1/2} = \sum_{j=1}^{N} \lambda_j\, \underline{u}_j\, \underline{u}_j^T.$$

Designating the dyadic matrix $\underline{u}_j\, \underline{u}_j^T$ by J_j, we have $J_j J_k = 0$, and $J_j^2 = J_j$ from the orthonormality of $\{\underline{u}_j\}$. Then $e_D^{1/2}\, a^k\, e_D^{-1/2} = \Sigma\, \lambda_j^k\, J_j$, so that

$$e_D\, a^k = \sum_j \lambda_j^k\, \underline{w}_j\, \underline{w}_j^T, \tag{3.2.2}$$

with

$$\underline{w}_j = e_D^{1/2}\, \underline{u}_j.$$

Next, let $N(t)$ be a time-reversible ergodic chain in continuous time, governed by $\{\nu, a_\nu\}$ as in §2.1. Then

$$e_D\, p(t) = \sum_{k=0}^{\infty} e^{-\nu t} \frac{(\nu t)^k}{k!}\, e_D\, a_\nu^k = \tag{3.2.3}$$

$$\sum_{j=1}^{N} \sum_{k=0}^{\infty} e^{-\nu t} \frac{(\nu t)^k}{k!}\, \lambda_j^k\, \underline{w}_j\, \underline{w}_j^T = \sum_{j=1}^{N} e^{-\nu t(1-\lambda_j)}\, \underline{w}_j\, \underline{w}_j^T.$$

In particular, for $m \in \mathscr{N}$,

$$e_m\, p_{mm}(t) = \sum_{j=1}^{N} e^{-\nu t(1-\lambda_j)} (w_{jm})^2 = \sum_{j=1}^{N} p_j\, e^{-\alpha_j t},$$

with

$$\alpha_1 = 0,\ \alpha_j > 0,\ \ j = 2,\dots,N,\ \ p_j \geq 0,\ \Sigma\, p_j = e_m.$$

Hence $p_{mm}(t)$ is a mixture of exponentials, i.e., is completely monotone (see Feller II, [15]) and has the appearance shown below. Complete monotonicity is an important property,

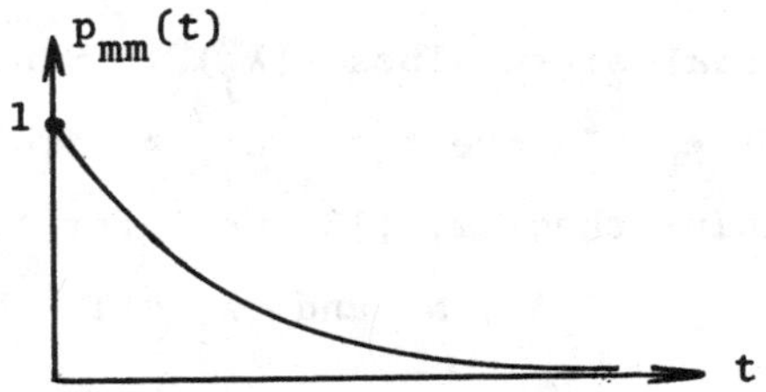

as we will see subsequently.

From (3.2.1), (3.2.2) and (3.2.3) we see easily that $P(t)$ has the spectral representation

$$P(t) = \sum_{j=1}^{N} e^{-\alpha_j t} (e_D^{-1/2}\underline{u}_j)\ (\underline{u}_j^T\ e_D^{1/2}) = \sum_{j=1}^{N} e^{-\alpha_j t}\ J_j, \tag{3.2.4}$$

where J_j is a matrix of rank one (a "dyad"), and

$$J_j\, J_k = 0 \quad j \neq k \qquad J_j^2 = J_j. \tag{3.2.5}$$

The dyad matrices J_j are therefore idempotent and orthogonal. For this ergodic case, the principal value of α_j is $\alpha_1 = 0$, and the corresponding principal dyad is $J_1 = \underline{1}\ \underline{e}^T$. In component form the spectral representation (3.2.4) becomes

$$p_{mn}(t) = \sqrt{\frac{e_n}{e_m}} \sum_{j=1}^{N} e^{-\alpha_j t} u_m^{(j)} u_n^{(j)}, \tag{3.2.6}$$

where $\alpha_1 = 0$; $\alpha_j > 0$, $j \neq 1$; $u_m^{(1)} = \sqrt{e_m}$; the vectors $\underline{u}^{(j)}$

are orthonormal; $\sum_{j=1}^{N} [u_m^{(j)}]^2 = 1$, all m; and all $u_m^{(j)}$ are real.

§3.3. Potentials; Spectral Representation.

In classical potential theory [43] arising from the study of gravitational and electrostatic fields (the reader need not known anything about these topics in physics) a certain property of path independence underlies the availability of the field potential. We will see that a similar path independence is present for time-reversible chains.

In a time-reversible ergodic chain governed by $a = (a_{mn})$, let

$$\theta_{mn} \stackrel{\text{def}}{=} \frac{a_{mn}}{a_{nm}} \tag{3.3.1}$$

for all directly communicating states. We recall (§2.4) that $a_{mn} > 0$ implies $a_{nm} > 0$ for such chains, so that θ_{mn} is meaningfully defined, and $\theta_{mn} = e_n/e_m$. Consider a path p in state space $\mathcal{N}$, $p = \{n_1, n_2, \ldots, n_k\}$ that is realizable in the sense that $a_{n_1 n_q} a_{n_2 n_3} \ldots a_{n_{k-1} n_k} > 0$. Then the path product

$$\prod_p \theta_{mn} = \theta_{n_1 n_2} \theta_{n_2 n_3} \ldots \theta_{n_{k-1} n_k} = \frac{e_{n_2}}{e_{n_1}} \frac{e_{n_3}}{e_{n_2}} \ldots \frac{e_{n_k}}{e_{n_{k-1}}} = \frac{e_{n_k}}{e_{n_1}},$$

is independent of the path chosen between any pair of designated states. In classical potential theory, the essence of a potential is the availability for a vector field $\underline{\varepsilon}(\underline{r})$ a function $V(\underline{r})$, such that

$$\int_c \underline{\varepsilon}(\underline{r}) \cdot d\underline{r} = V(P_0) - V(P_T),$$

independent of path c between P_0 and P_T. In such a case

$\underline{\varepsilon} = -\nabla V$. The path product (3.3.2) can be rewritten in terms of a sum, i.e., $\prod_p \theta_{mn} = \exp(\sum_1^{k-1} \log \theta_{n_j n_j+1})$, and the analogy with (3.3.2) is clear.

Path independence permits *extension of reversibility in time to transient chains*. We next introduce potential coefficients. Let $\mathcal{N}$ be finite and let $\mathbf{a} = (a_{mn})$ govern N_k. Suppose $\prod_{j=1}^{k-1} \theta_{n_j n_{j+1}}$ depends on n_1 and n_k only and not on the path between n_1 and n_k. We now drop the assumption of ergodicity, keeping the state space finite. For this we allow permanent loss from the state space.† Designate some state $n_0 \in \mathcal{N}$, put $\pi_{n_0} = 1$ and evaluate the coefficient

$$\pi_n = \pi_{n_0} \theta_{n_0 n_1} \cdots \theta_{n_{k-1} n}, \quad n \in \mathcal{N}, \tag{3.3.3}$$

for some arbitrary path $p = \{n_0, n_1, \ldots, n_{k-1}, n\}$. Note that π_n is unique because of path independence. The coefficients π_n are called *potential coefficients*, and are positive. From (3.3.3), it follows that $\pi_m a_{mn} = \pi_n a_{nm}$ (from the definition of θ_{mn}), for all $m, n \in \mathcal{N}$. Hence $\pi_D \mathbf{a}$ is symmetric, where π_D is a diagonal matrix with positive elements on the diagonal.

The path independence and resultant symmetry of $\pi_D \mathbf{a}$ implies that when $\mathbf{a}$ is ergodic, the potential coefficients and ergodic probabilities differ only by a multiplicative

---

†The situation is clearest in discrete time. Suppose we have a chain on a state space $\mathcal{N}$ where one state a is absorbing and all other states in the set $\mathcal{N}$-{a} communicate. If the process starts in $\mathcal{N}$- {a}, only the transition probabilities between its states are of importance, and the behavior of the system can be discussed using the strictly substochastic matrix $\mathbf{a}^*$ involving these. The process on $\mathcal{N}$-{a} may be thought of as a "lossy" Markov chain. This point of view is helpful, even if not in keeping with convention.

normalization constant. Specifically we have

$$e_n = \frac{\pi_n}{\Sigma\ \pi_n}. \tag{3.3.4}$$

To see this we note that $\pi_D a$ symmetric implies the symmetry of $\pi_D a^k$ as in §1.3. Hence, when a is ergodic, the limit $\lim \pi_D a^k = \pi_D \underline{1}\ \underline{e}^T$ is symmetric. It follows that $\pi_m e_n = \pi_n e_m$ so that $e_n = e_0 \frac{\pi_n}{\pi_0} = e_0 \pi_n$ and the ergodic distribution is found by normalizing the potentials. It follows from (3.3.3) and (3.3.4) that the ergodic probabilities are obtainable from the path products and are therefore highly accessible. Note from (3.3.3) (3.3.1) and (2.1.3) that

$$\pi_n = \pi_{n_0} \frac{\nu_{n_0 n_1}}{\nu_{n_1 n_0}} \frac{\nu_{n_1 n_2}}{\nu_{n_2 n_1}} \cdots \frac{\nu_{n_{k-1} n_k}}{\nu_{n_k\ n_{k-1}}}, \tag{3.3.5}$$

for any realizable path with $n_k = n$.

Two other points are in order:

1. When $p(t) = e^{-\nu t[I-a_\nu]}$ and a_ν is substochastic, corresponding to loss from the state space of interest, $\underline{\pi}$ is not in general an eigenvector of a_ν and $p(t)$. Only for the case of uniform loss from each state is this possible. For $(\underline{\pi}^T a)_n = \Sigma_m \pi_m a_{mn} = \pi_n \Sigma_m a_{nm} = \lambda \pi_n$ requires $\Sigma_m a_{nm} = \lambda \leq 1$ for all n, i.e., uniform loss.

2. When $p(t)$ and a_ν are substochastic, the matrix $p(t)$ still has a spectral representation of the form (3.2.4), i.e., we will have

$$p(t) = \sum_{j=1}^{N} e^{-\alpha_j t} J_j, \tag{3.3.6}$$

where the J_j are real orthogonal and idempotent as in (3.2.5). When loss is present, the principal value α_1 is positive, and the principal dyad has the form

$$(J_1)_{mn} = \frac{r_m \ell_n}{\sum_k r_k \ell_k} . \tag{3.3.7}$$

For the ergodic case r_m is independent of m. When loss is present, all values α_j are positive. For the components one has

$$p_{mn}(t) = \sqrt{\frac{\pi_n}{\pi_m}} \sum_{j=1}^{N} e^{-\alpha_j t} u_m^{(j)} u_n^{(j)} , \tag{3.3.8}$$

with $\underline{u}^{(j)}$ orthonormal eigenvectors of $\pi_D^{1/2} a \pi_D^{-1/2}$. In particular, we note that the diagonal terms are completely monotone, i.e.,

$$p_{mm}(t) = \sum_{j=1}^{N} e^{-\alpha_j t} u_m^{(j)2} .$$

§3.4. More General Time-Reversible Chains.

We are now in a position to extend the notion of time-reversibility to chains other than ergodic, and to permit, in particular, system loss.

Definition 3.4A. Any Markov chain $N(t)$ with path independence for $\prod_p \frac{\nu_{mn}}{\nu_{nm}}$ as in §3.3 will be called reversible in time.

Again we emphasize the availability of the spectral representation (3.3.6) for such chains.

§3.5. Process Modifications Preserving Reversibility.

A time-reversible process in the sense of definition 3.4A can be altered in a variety of simple ways without destroying the reversibility. As we will see, the modified processes so obtained are of interest. Indeed, such a modified process may be of primary interest, and the unmodified process may be a construct which is a stepping stone to the exploration of the primary process.

A. Censored transitions.

Let N_k on state space $\mathscr{N}$ be governed by $\mathbf{a}$, ergodic and time-reversible. Let S be a subset of $\mathscr{N}$, and let all transitions to S be censored, i.e., for any $m,n \in \mathscr{N}-S$, put $a^*_{mn} = a_{mn}$, $m \neq n$; $a^*_{mm} = a_{mm} + \Sigma_{k \in S}\, a_{mk}$. The matrix $\mathbf{a}^*$ is stochastic. Consider the process N^*_k governed by $\mathbf{a}^*$ restricted to the state space $\mathscr{N}$-S. Then N^*_k is still reversible in time, since the path independence of $\Pi\, \theta_{mn}$ is unaffected on $\mathscr{N}$-S. Let the state 0 be in $\mathscr{N}$-S and let the potential coefficient π_0 be held fixed. If the censoring does not destroy irreducibility N^*_k is ergodic, and $\pi_n = \pi_0 \prod_p \theta_{n_j n_{j+1}}$ is unaltered for feasible paths p, i.e., the potential coefficients are unchanged. Hence $\pi_D \mathbf{a}^*$ is symmetric, and automatically $\pi_D \mathbf{a}^{*k}$ is symmetric. It follows that the limit $\pi_D \underline{1}\, \underline{e}^{*T}$ is also symmetric. Hence $\pi_n e^*_m = \pi_m e^*_n$, and the new ergodic distribution is obtained by renormalizing the potentials, or equivalently the old ergodic probabilities, on $\mathscr{N}$-S. An example of this instance is provided in §2.5, under C, where $S = \{\underline{n}\colon n_1 + n_2 > L\}$. The joint process $[N_1(t), N_2(t)]$ where $N_1(t)$ and $N_2(t)$ are independent birth-death processes does not lose its reversibility when the constraint $n_1+n_2 \leq L$ is imposed. Its ergodic probabilities are then

$$e^*_{n_1,n_2} = \frac{e^{(N_1)}_{n_1}\, e^{(N_2)}_{n_2}}{\sum_{n_1+n_2 \leq L} e^{(N_1)}_{n_1}\, e^{(N_2)}_{n_2}} \,,$$

a result by no means transparent. Insight into the transient behavior of the modified process is also available through the reversibility, as will be seen.

The reader will observe that a state need not be eliminated under censoring. If two states A and B have direct transition rates ν_{AB} and ν_{BA}, altering the process by setting both these rates equal to zero does not destroy the reversibility if the irreducibility is still intact.

B. <u>States made absorbing</u>.

To study the transient behavior of an ergodic process, it is often helpful to modify the process by making certain states absorbing. If, for example, one were interested in the passage time density $s_{L,L+1}(\tau)$ from state L to L+1 of an ergodic birth-death process, one could make state L+1 absorbing, start at state L and evaluate $\sum_0^L p_n^*(t)$, say.

Let $T \subset \mathcal{N}$ be made *absorbing* and consider the process N_k^* on $\mathcal{N}$-T. Again $\prod_p \theta_{mn}$ are unaltered for paths within $\mathcal{N}$-T, so that $\pi_D \mathfrak{a}$ *is symmetric* on $\mathcal{N}$-T and a spectral representation can be given for transitions of the transient chain N_k^*. For our *birth-death process* let the transition rates be $\{\lambda_n, \mu_n\}_{n=0}^{\infty}$, $\mu_0 = 0$, $\lambda_0 > 0$, $\lambda_n, \mu_n > 0$, $n = 1,2,\ldots$ Take $T = \{n: n > L\}$. Then the birth-death process on $\mathcal{N}$-T is transient. As in §3.2, we get from the spectral representation (3.3.9) of $\mathfrak{p}(t)$, that $p_{mm}^*(t) = \sum_j \gamma_m^{(j)} e^{-\alpha_j t}$ with $\alpha_j > 0$, $\gamma_m^{(j)} > 0$. Applying this with $m = L$, we find that the passage time density $S_{L,L+1}(t)$ from state L to L+1 is completely monotone, as follows:

$$s_{L,L+1}(t) = \lambda_L \, p_{LL}^*(t) = \lambda_L \Sigma \, \gamma_L^{(j)} \, e^{-\alpha_j t}.$$

A technical extension of the argument shows that in a birth-death process, any passage time to a neighboring state has completely monotone density.† Indeed the reader will verify in the same way that for any ergodic tree process (2.5), passage times to adjacent states have completely monotone densities.

The reader will also verify that any time-reversible process modified by introducing loss rates from any state to the exterior of the state space is still time-reversible.

C. Aggregation of states.

Let $N(t)$ be an ergodic reversible process on state space $\mathscr{N}$. Let $\mathscr{S} \subset N$. The states of $\mathscr{S}$ may be aggregated to form a new state S and a new process $N^*(t)$ in the following way. Let

$$\begin{aligned} \nu^*_{mn} &= \nu_{mn} \\ \nu^*_{mS} &= \sum_{\mathscr{S}} \nu_{ms} \\ e_{S*} &= \sum_{\mathscr{S}} e_s \\ \nu^*_{Sm} &= \frac{e_m}{e^*_S}\, \nu^*_{mS} = \frac{e_m \sum_{\mathscr{S}} \nu_{ms}}{\sum_{\mathscr{S}} e_s}\,, \end{aligned} \tag{3.5.1}$$

for all m,n not in $\mathscr{S}$. Then, clearly, the new process has detailed balance, and is reversible in time.

§3.6. Replacement Processes.

Let N_k be ergodic on state space $\mathscr{N}$, with steps governed by $\mathfrak{b}$. Let the state space $\mathscr{N}$ be partitioned into two sets G and B (for good and bad), i.e. let $\mathscr{N} = G+B$,

†cf. Chapter 5.

where G and B are disjoint. Suppose that the process starts in G, and that whenever a process sample path arrives at B, it is replaced at a particular state r in G, the replacement state. The process is then called a replacement process. For the process N_k^* on G, the governing matrix b^* has components

$$b^*_{mn} = b_{mn} + (\sum_B b_{ms})\, \delta_{nr} \qquad m,n \in G. \tag{3.6.1}$$

Such a replacement process is not in general reversible, since detailed balance is violated.

An important example of a replacement process is that obtained from systems of independent Markov components of interest to reliability theory discussed in §2.5. When certain critical system states are reached, a maintenance policy may specify that all components that are not working be immediately replaced. Such a policy replaces system process samples reaching the critical states at the perfect state with all components up. Replacement processes will be discussed further in the next chapter.

Chapter 4

Potential Theory, Replacement, and Compensation

§4.00.

The potential for a transient chain N_k governed by transition matrix $\mathfrak{a}$ is the matrix $\mathfrak{g} = \sum_0^\infty \mathfrak{a}^k$. The corresponding potential for transient chains in continuous time is $\mathfrak{g}(t) = \int_0^\infty \mathfrak{p}(t)\, dt$ where $\mathfrak{p}(t)$ is the transition matrix.† These potentials appear as entities in the answers to many important questions associated with a particular chain, ergocic or transient. For example, certain replacement processes $N^*(t)$ arise from a chain $N(t)$ when samples reaching a bad set B are replaced at a particular "replacement state". The ergodic distribution for the replacement process involves a potential obtained from $N(t)$ in a simple way. Directly related to the idea of replacement is a method of compensation which treats a modified process as the original process altered by the insertion of positive and negative mass at "boundary states" in such a way as to generate the modified process. In this way structural simplicity of the original process due to spatial homogeneity or independence can be

† Other remarks on potential theory in an ergodic setting are given in §7.00.

retained in the treatment of the modified process. A variety of examples are given.

§4.1. The Green Potential.

Probabilistic potential theory [61] centers about the structure of the matrix

$$\mathfrak{g} = \sum_{0}^{\infty} \mathfrak{a}^k, \tag{4.1.1}$$

where $\mathfrak{a}$ is the single-step transition probability matrix of a Markov chain, typically transient on a denumerably infinite state space. No knowledge of potential theory is required of the reader.

Our interests have a much simpler setting, a finite state space and a lossy chain, so that the matrix $\mathfrak{a}$ in (4.1.1) is strictly substochastic. We will see that the matrix $\mathfrak{g}$, which we will call the Green potential, appears intact in a variety of contexts. For our finite lossy chains one may write

$$\mathfrak{g} = [\mathfrak{I}-\mathfrak{a}]^{-1}. \tag{4.1.2}$$

For a lossy chain $N(t)$ in continuous time on a finite state space governed by hazard rates $\{\nu_{mn}\}$, the uniformization bridge of §2.1 implies that $N(t)$ is equivalent in law to a finite chain with $\mathfrak{a}_\nu$ and an independent Poisson process $K(t)$ of rate ν. Then

$$\begin{aligned} \mathfrak{p}(t) &= \exp\,[-\nu t(\mathfrak{I}-\mathfrak{a}_\nu)] \\ &= \sum_{k=0}^{\infty} e^{-\nu t}\,\frac{(\nu t)^k}{k!}\,\mathfrak{a}_\nu^k. \end{aligned} \tag{4.1.3}$$

Hence

$$\int_0^\infty \mathfrak{p}(t)\,dt = \sum_{k=0}^{\infty} \{\int_0^\infty e^{-\nu t} \frac{(\nu t)^k}{k!}\,dt\}\mathfrak{a}_\nu^{\,k} = \frac{1}{\nu}\sum_0^\infty \mathfrak{a}_\nu^{\,k}, \tag{4.1.4}$$

for all permissible ν. For chains in continuous time the role of the Green potential is played by $\int_0^\infty \mathfrak{p}(t)\,dt$. In this chapter we will again be treating discrete time and continuous time chains simultaneously without specific mention.

§4.2. The Ergodic Distribution for a Replacement Process.

Consider the replacement process N_k^* of §3.6 on the state space $\mathcal{N}^* = G = \mathcal{N}-B$ with replacement at state r. Let $\mathfrak{b}_G$ be the transition matrix of the origional process N_k truncated to G, i.e., let $(\mathfrak{b}_G)_{mn} = b_{mn}$ for $m,n \in G$. Then $\mathfrak{b}_G$ is strictly substochastic. Let the state probability vector for N_k^* be denoted by $\underline{p}_k^*$ with components $(p_k^*)_n = P[N_k^* = n]$, for $n \in G$. Let the replacement process start at the replacement state r with $\underline{p}_0^* = \hat{\underline{u}}_r = (0,0,..,1,...0)$, the r'th unit vector.

Further, let $h_k = P\{\text{replacement at epoch } k\}$. Then

$$\underline{p}_k^{*T} = \hat{\underline{u}}_r^T (h_k I + h_{k-1}\mathfrak{b}_G + h_{k-2}\mathfrak{b}_G^2 + \dots + \mathfrak{b}_G^k) = \hat{\underline{u}}_r^T \sum_{m=0}^{k} h_{k-m}\mathfrak{b}_G^m. \tag{4.2.1}$$

From renewal theory, if the mean time between replacements is $\mu < \infty$, we know that $h_k \to \frac{1}{\mu}$ $(k \to \infty)$. Then, as shown in the next theorem,

$$\underline{p}_k^{*T} \to \frac{\hat{\underline{u}}_r^T \Sigma_o^\infty \mathfrak{b}_G^m}{\mu} = \frac{\hat{\underline{u}}_r^T \mathfrak{g}_G}{\mu} = \underline{e}^{*T}, \tag{4.2.2}$$

where $\mathfrak{g}_G$ is the Green potential of $\mathfrak{b}_G$. Here then, is the first instance of the appearance of the Green potential. We note that

$$\mu = \underline{\hat{u}}_r^T \mathfrak{g}_G \underline{1}. \tag{4.2.3}$$

Theorem 4.2A. If for the replacement process N_k^*, the replacement time τ has expectation $E\tau = \mu < \infty$ and N_k^* is an irreducible process on G, then N_k^* is ergodic and the ergodic probabilities e_n^* are given by $\underline{e}^* = \underline{\hat{u}}_r^T \mathfrak{g}_G/\mu$.

Proof: Let P_k denote the probability that no replacement has taken place up to and including the k'th step. Then $P_k = \underline{\hat{u}}_r^T \mathfrak{b}_G^k \underline{1}$, and $\mu = (P_0 - P_1) + 2(P_1 - P_2) + \ldots = P_0 + P_1 + \ldots = \underline{\hat{u}}_r^T \Sigma_o^\infty \mathfrak{b}_G^k \underline{1} < \infty$ by assumption. Since $h_k = P_{k-1} - P_k$ is bounded, one may use the dominated convergence theorem in (4.2.1) for $k \to \infty$, to obtain Theorem (4.2A).

Remark 4.2B. Replacement may be at one of several states in G with probabilisties α_r, that $r \in G$ is chosen as replacement state. In that case the ergodic probabilities are a mixture of probabilities of the form (4.2.2) as is easily seen. One gets $\underline{e}^{*T} = \Sigma_r \alpha_r \underline{\hat{u}}_r^T \mathfrak{g}_G/(\Sigma \alpha_r \mu_r)$ where u_r is the mean passage time from replacement state r to the bad set B. One may also write $\underline{e}^{*G} = \underline{h}^T \mathfrak{g}_G$, with $\underline{h}^T$ the "replacement vector", with components $\alpha_r/\Sigma \alpha_r \mu_r$ $(r \in G)$.

Example 4.2C. Suppose the behavior of an *inventory system* is modeled by a modified homogeneous birth-death process, i.e., there is a probability λ per unit time that an item arrives, η that an item is removed. The modification is that at depletion of the inventory, there is restocking to level $L > 0$. In this setting $B = \{0\}$ and the ergodic distribution on $G = \{1,2,\ldots,L, L+1,\ldots\}$ is given by $\underline{e}^{*T} = \underline{\hat{u}}_L^T \mathfrak{g}_G/\mu$, where

μ is the mean passage time from L to 0. The reader may note that if $\mathcal{E}_G$ is available, μ may be obtained by normalization.

§4.3. The Compensation Method.

For the study of modified processes or of transient processes associated with a specified primary process, a method of compensation generalizing the procedure in §4.2 is very useful. For the compensation method one retains the structural simplicity of the primary process arising from its spatial homogeneity or independence, by working with a potential or transition probability function natural to the primary process. The primary process is modified to the desired process via certain flow sources inserted at the boundary states of the state space at which the transitions of the primary process are altered. All this will become clear from examples.

For the first example, we will derive the ergodic distribution for a spatially homogeneous random walk, modified by boundaries, and see that this ergodic distribution is given by the convolution of the potential of the homogeneous process, and a compensation measure.

§4.4. Notation for the Homogeneous Random Walk.

Notation will be defined concurrently for discrete time and continuous time processes. For discrete time, let $X_k = X_0 + \Sigma_1^k \xi_j$, with ξ_j i.i.d. random variables, or random vectors, as the case may be. Let the increment measure† be

†The notation of measure theory is powerful and succinct, but may be unfamiliar to many readers. It is easy to paraphrase the results for increments having probability density functions or probability masses on the lattice of integers. The reader is urged to do so.

$a(A) = P\{\xi \in A\}$, and let the initial distribution be $f_0(A) = P\{X_0 \in A\}$. Then, one has

$$f_k(A) \stackrel{\text{def}}{=\!=} P\{X_k \in A\} = f_0 * a^{(k)}(A). \tag{4.4.1}$$

In continuous time, there exists a Poisson process $K_\nu(t)$ with rate ν, indicating the number of increments in $[0,t]$ corresponding to a constant hazard rate for the occurrence of an increment (cf. §2.2). One then has

$$X(t) = X_0 + \sum_0^{K_\nu(t)} \xi_j . \tag{4.4.2}$$

Let the k-fold convolution of the increment distribution be denoted by g_k, i.e.,

$$g_k(A) = a^{(k)}(A). \tag{4.4.3}$$

Correspondingly, in continuous time one has the *time-dependent Green measure* $g(t,A)$, defined by

$$g(t,A) = P\{X_t \in A | X_0 = \underline{0}\} = \sum_0^\infty e^{-\nu t} \frac{(\nu t)^k}{k!} a^{(k)}(A). \tag{4.4.4}$$

A second kind of Green measure, with the character of a potential, is the *ergodic Green measure*, defined in discrete time as

$$g(A) \stackrel{d}{=} \sum_0^\infty a^{(k)}(A), \tag{4.4.5}$$

which must be finite, and will be assumed finite for the moment. For transient chains on the lattice the result is immediate (see §4.7). If the process is one-dimensional,

$$g((-\infty,x]) = U(x) + \sum_1^\infty A^{(k)}(x) \tag{4.4.6}$$

where $U(x)$ is the unit step function with $U(x) = 1$ for $x \geq 0$, 0 for $x < 0$, and $A(x)$ is the c.d.f. corresponding to the measure a. It is known that [23], [60] $g(A) < \infty$ for all bounded (compact) A, if $E\xi \neq 0$, $E|\xi| < \infty$, and $g(A) = \int_0^\infty g(t,A)\,dt$.

§4.5. The Compensation Method Applied to the Homogeneous Random Walk Modified by Boundaries.

A formal treatment of the method is followed by an intuitive interpretation of the procedure. The reader is urged to skim the formal procedure, go to the interpretation, and work back and forth between them.

Consider a spatially (and temporally) homogeneous process modified by boundaries. An example is the *Lindley process*, defined by

$$X_{k+1} = \max\,[0,\ X_k + \xi_{k+1}], \tag{4.5.1}$$

where ξ_j is a sequence of i.i.d. random variables. In this process, whenever a virtual transition is into the negative half-line, the modified process law resets the value of the process at 0. This process first studied by Lindley [52], describes the waiting time before service of successive customers to a single server when the service times are i.i.d., and the interarrival times are i.i.d., and the discipline is first-come-first-served. One can also consider Lindley-type processes on the positive quadrant in R_2, with any of several boundary replacement rules.

Let X_k be governed by transition kernel $a(x,A) = P\{X_{k+1} \in A | X_k = x\}$. Then $f_k(A) \overset{\text{def}}{=\!=} P\{X_k \in A\}$ satisfies

$$f_{k+1}(A) = \int a(x,A)\, f_k(dx). \tag{4.5.2}$$

As in (3.6.1), the kernel of the modified process will be

related to the homogeneous one as follows. One writes $a(x,A)$ in the form

$$a(x,A) = a_H(x,A) + D(x,A), \tag{4.5.3}$$

with $a_H(x,A) = a_0(A-x)$, where $a_0(A) = P[\xi \in A]$; and $D(x,A)$ is defined by (4.5.3). Clearly $D(x,A)$ is a signed measure, bounded by 1, and with total mass 0. Substituting (4.5.3) into (4.5.2) one obtains

$$f_{k+1}(A) = a_0 * f_k(A) + c_{k+1}(A), \tag{4.5.4}$$

where

$$c_{k+1}(A) = \int D(x,A)\, f_k(dx) = \int a(x,A) f_k(dx) - \int a_H(x,A)\, f_k(dx). \tag{4.5.5}$$

This says that the compensation measure is the difference between the real arrival rate at time k for the modified process and the virtual arrival rate for the process if the homogeneous transition laws were operative. If one defines $c_0(A) = f_0(A)$, (4.5.4) is formally solved by

$$f_k(A) = \sum_{k'=0}^{k} a_0^{(k-k')}(A) * c_{k'}(A), \tag{4.5.6}$$

where $a_0^{(0)}(A) = \delta(A) = 1$ if $0 \in A$, 0 otherwise. If the underlying process is ergodic, then f_k will converge to the ergodic measure f_∞, and hence, from (4.5.5), c_k will converge to a limit measure c_∞. Then, as in Theorem 4.2A, from (4.5.6) one obtains

$$f_\infty(A) = g(A) * c_\infty(A). \tag{4.5.7}$$

Here, $g(A)$ is the homogeneous potential (4.4.5) and

$$c_\infty(A) = \int D(x,A)\, f_\infty(dx), \tag{4.5.8}$$

where $g(A)$ is finite.

Let us now take stock. Equation (4.5.7) states that the *ergodic measure is the homogeneous potential of the compensation measure.* The character of $c_k(A)$ and $c_\infty(A)$ as compensation may be seen in equation (4.5.4) where $c_{k+1}(A)$ acts as a correction term for the homogeneous process made necessary by the boundary. The reader will note from (4.5.3) and (4.5.5) that the compensation measure is localized at the boundary.

An intuitive understanding of the compensation measure may be obtained in the following way. Suppose the homogeneous process is modified by an impenetrable boundary as in our Lindley process. The modification may be visualized as neutralizing process samples entering the set $-\infty < n < 0$ by placing negative mass at the arrival states to the precise extent necessary to have a net zero arrival rate there, and simultaneously inserting positive mass at state $n = 0$ as required for the modified process. The compensation measure represents the neutralizing mass and replacement mass. The situation is made more vivid in a lattice example in the next section. The procedure works when the homogeneous process is transient so that $g(A)$ is finite. We have yet to convince the reader that the procedure is worthwhile.

4.6. Advantages of the Compensation Method. An Illustrative Example.

The compensation method has two important advantages:

1. It enables the modified process to be discussed in terms of the potential of the homogeneous process, where the greater simplicity present often enables one to calculate this

potential fairly easily.

2. The procedure enables one to convert the problem of finding the ergodic distribution of the modified process to that of finding the ergodic compensation measure. This is often far easier because of the structure of the increment measure which makes the second problem one of much lower rank. This will be clear from our example.

3. The classical method of solving the equation for the Lindley process is a method in the complex plane called the Wiener-Hopf method or Hilbert method, a terrifying procedure to a complex plane novice. Moreover, the Wiener-Hopf method requires the location of zeros in the complex plane of certain equations which may be hopeless. The compensation method is a method in the real domain, more readily understood and more subject to the guidance of human intuition. The search for complex zeros is replaced by the solution of a set of simultaneous linear equations. More information may be found in [23], where many examples are given.

Example 4.6A. Consider a skipfree-negative homogeneous Markov chain in discrete time, on the non-negative integers, i.e., $X_k = X_0 + \Sigma\, \xi_j$, where the ξ_j are i.i.d. integers such that $1+\xi \geq 0$ with probability one. Let $a_n = P[\xi_j = n]$. Let $f_n^{(k)} = P[X_k = n]$ and let $c_k(A)$ have mass $c_n^{(k)}$ at state n. Then (4.5.5) takes the form $c_{-1}^{(k+1)} = -a_{-1}\, f_0^{(k)}$; $c_0^{(k+1)} = a_{-1}\, f_0^{(k)}$ with all other compensation mass zero. Hence $c_\infty(A)$ has support on $\{-1,0\}$ with $c_{-1} = -c$, $c_0 = +c$ for some $c > 0$, as shown below. From (4.5.7) the

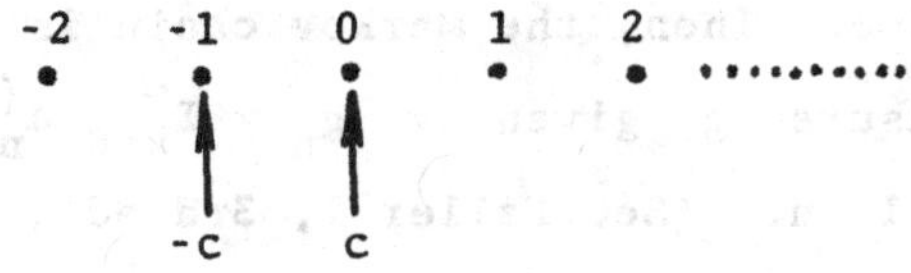

ergodic distribution is given by

$$e_n = c(g_n - g_{n+1}), \tag{4.6.1}$$

with

$$g_n = \sum_0^{\infty} a_n^{(k)}. \tag{4.6.2}$$

If $\mu = \Sigma_{-1}^{\infty} n\, a_n$ exists and is negative, $g_n < \infty$ for all n. The constant c may be found by renormalizing. If for this lattice problem one had $\xi \geq -k$, the compensation mass would have negative support at $\{-k, -k+1 \ldots -1\}$ and counterbalancing positive support at zero. The unknown compensation masses would be found from the requirement that $e_n = 0$ for $-k \leq n \leq -1$ and $\Sigma\, e_n = 1$. This gives the set of equations

$$\sum_{-k}^{0} c_m\, g_{n-m} = 0, \quad -k \leq n \leq -1$$

which with normalization gives c_m. The rank reduction advantages are then clear.

§4.7. Exploitation of the Structure of the Green Potential for the Homogeneous Random Walk.

Besides its rank reducing properties, the compensation method has a second advantage, in that in several cases of interest the Green measure has simple properties, which can be exploited, as the following will show.

Example 4.7A. Consider a random walk on the lattice of integers, with increment distribution $a_n = P\{\xi = n\}$, $n \in Z$. For definitions and notation, refer back to §4.4. Let $E\xi$ exist and be *negative*. Then, the Markov chain is transient, and the Green measure g, given by $g_n = \Sigma_{k=0}^{\infty} a_n^{(k)}$, exists and is finite for all n. (See Feller I, 3rd ed., p. 389.)

<u>Lemma 4.7B</u>. If ξ is skipfree-negative and $E\xi < 0$, then $g_n = g_0$ for $n \leq 0$. If ξ is skipfree-positive and $E\xi < 0$, then $g_n = \theta^n g_0$ for some $0 < \theta < 1$ and all $n \geq 0$.

<u>Proof</u>: A continuity of probability argument gives

$$p_{0n}^{(k)} = a_n^{(k)} = \sum_{k'=0}^{k} s_{0n}(k') p_{nn}(k-k') = s_{0n} * p_{nn}^{(k)}, \quad (4.7.1)$$

where $s_{0n}(k) = P\{X_k = n, X_j \neq n, 0 < j < k | X_0 = 0\}$, the first passage probability from 0 to n in k steps. From the spatial homogeneity of the random walk, one has $p_{nn}(k) = p_{00}(k)$, so that one gets from (4.7.1) by summation

$$g_n = \sum_{k=0}^{\infty} a_n^{(k)} = \sum_{k=0}^{\infty} \sum_{k'=0}^{k} s_{0n}(k') \, p_{00}(k-k'). \quad (4.7.2)$$

Interchanging the order of summation, one then obtains

$$g_n = \sum_{k'=0}^{\infty} s_{0n}(k') \sum_{j=0}^{\infty} p_{00}(j) = g_0 \sum_{k'=0}^{\infty} s_{0n}(k). \quad (4.7.3)$$

For a skipfree-negative process and negative n, the first passage time from 0 to n is an honest random variable (since $E\xi < 0$), and hence $\sum_{k=0}^{\infty} s_{0n}(k) = 1$. This proves the first part of the lemma.

If ξ is skipfree-positive, one writes for positive n, $s_{0n}(k) = s_{01} * s_{12} * s_{n-1\,n}(k)$, and using spatial homogeneity, one obtains

$$\sum_{k=0}^{\infty} s_{0n}(k) = \sum_{k=0}^{\infty} s_{01}^{(n)}(k) = \left(\sum_{k=0}^{\infty} s_{01}(s)\right)^n = \theta^n, \text{ with } 0 < \theta < 1.$$

Hence, from (4.7.3) $g_n = \theta^n g_0$.

Remark 4.7C. Both conclusions of the lemma hold if the process is skipfree in both directions. In that case the Green measure is known, except for two constants, g_0, θ. The form

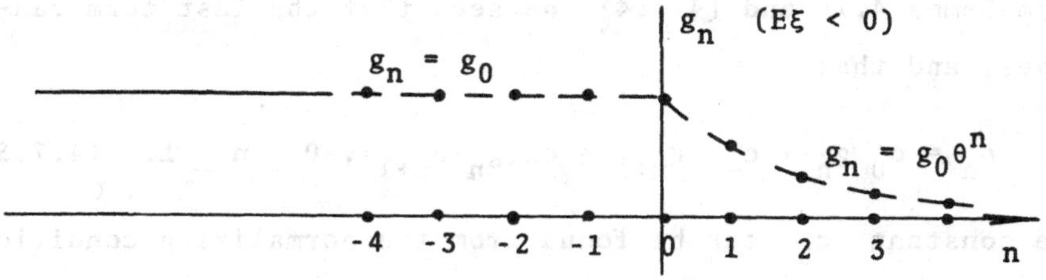

of g_n is shown in the figure. Lemma 4.7B carries over to processes on the continuum homogeneous in space and time, e.g., the Wiener process with drift. See: Sptizer, F. *Principles of Random Walk*, Van Nostrand, Princeton, N.J., 1964, or: Keilson, J. *Greens Function Methods*, Hafner, New York, 1965 (abbr. GFM henceforth).

We will now show how Lemma 4.7B may be exploited. Returning to Example 4.7A, suppose that a skipfree-negative random walk with $E\xi < 0$ is constrained by *two* retaining boundaries, at 0, and $L > 0$. The boundary is said to be retaining or impenetrable if process samples are stopped and held at the boundary until permissible increments occur.

The compensation measure here has mass at 0,-1 for the boundary at 0 (as in Example 4.2.4), and at L, L+1, L+2,... for the boundary at L. Looking hard at (4.5.4) and (4.5.8), one concludes that $c_L > 0$, $c_{L+j} \le 0$, $j = 1,2,...$ and from the fact that the total compensation mass is zero at both boundaries, one has

$$c_L = -\sum_{j=L+1}^{\infty} c_j, \quad c_0 = -c_1. \tag{4.7.4}$$

From (4.5.7), the ergodic probabilities are

$$e_n = \Sigma_{-\infty}^{\infty} c_j\, g_{n-j} = c_0\, g_n + c_{-1}\, g_{n+1} + \sum_{j=L}^{\infty} c_j\, g_{n-j},$$
$$0 \leq n \leq L.$$

From Lemma 4.7B and (4.7.4) one sees that the last term vanishes, and that

$$e_n = c_0\, g_n + c_{-1}\, g_{n+1} = c_0(g_n - g_{n+1}),\ 0 \leq n \leq L. \tag{4.7.5}$$

The constant c_0 can be found from the normalizing condition $\Sigma_0^L\, e_n = 1$. Comparing (4.7.5) with the result obtained in Example 4.6A, one concludes that the ergodic distribution for a skipfree-negative random walk with *two* boundaries is obtained from the *one* boundary case by renormalization to the appropriate set of states.

§4.8. Similar Situations.

a) The compensation method can also be used in non-Markovian settings to exploit skipfree behavior. See, for example, [25] where the ergodic queue length distribution for queueing systems with finite capacity is treated in this manner.

b) The Takacs virtual waiting time process, modified by a retaining upper bound, can be thought of as describing the content of a finite dam with constant outflow and instantaneous inflow at Poisson epochs. The relation between the ergodic distributions of this case and of the infinite dam is as above. See GFM [23].

Chapter 5

Passage Time Densities in Birth–Death Processes; Distribution Structure

§5.00.

Birth-death processes are a simple family of time-reversible processes of great importance in applications, and their passage time densities are of corresponding interest. Two basic distribution forms appear in the study of densities, complete monotonicity and the PF_∞ form, that of the Polya-frequency densities of infinite order. These forms are special cases of log-concavity and log-convexity, each of interest in its own right. Log-concavity is a property closely related to unimodality, and is equivalent to "strong unimodality". The origion of these forms, related moment inequalities, and their relation to the IFR and DFR classes of reliability theory are discussed.

§5.1. Passage Time Densities for Birth-Death Processes.

A simple chain of great importance is the birth-death process. Consider a birth-death process with upward and downward hazard rates λ_n, μ_n, respectively, $\lambda_n > 0$, $n \geq 0$; $\mu_n > 0$, $n \geq 1$, $\mu_0 = 0$. Denote by $s_{0n}(\tau)$ the passage time density from 0 to n:

$s_{0n}(\tau) = -\frac{d}{d\tau} P\{X(t) < n,\ 0 < t < \tau \mid X(0) = 0\}$. The corresponding random variable is denoted by T_{0n}. Similarly, let $s_n^+(\tau) = s_{n\,n+1}(\tau)$ be the upward passage time density at n. Then

$$s_{0n}(\tau) = s_0^+ * s_1^+ * \dots * s_{n-1}^+(\tau) \tag{5.1.1}$$

or in random variables $T_{0n} = T_{01} + T_{12} + \dots + T_{n-1n}$, where the r.v.'s in the right side are independent. A recursive probabilistic argument obtains $s_{0n}(\tau)$ as follows. Let $\nu_n = \lambda_n + \mu_n$. The dwell time at state n has density $\nu_n e^{-\nu_n t}$, and with probability $\frac{\lambda_n}{\nu_n}$ there is a subsequent transition to $n+1$, with probability $\frac{\mu_n}{\nu_n}$ the transition is to $n-1$. Hence

$$s_n^+(\tau) = \frac{\lambda_n}{\nu_n} \nu_n e^{-\nu_n \tau} + \frac{\mu_n}{\nu_n} \nu_n e^{-\nu_n \tau} * s_{n-1}^+(\tau) * s_n^+(\tau), \quad n \geq 1. \tag{5.1.2}$$

The recursion is obtained, apart from inversion difficulties, by using Laplace transforms. Laplace transforms will be denoted by corresponding lower case Greek letters. From (5.1.2) one finds

$$\sigma_n^+(s) = \frac{\lambda_n}{s+\lambda_n+\mu_n - \mu_n \sigma_{n-1}^+(s)}, \quad n \geq 1. \tag{5.1.3}$$

One then obtains $\sigma_n^+(s)$ iteratively from (5.1.3) and

$$\sigma_0^+(s) = \frac{\lambda_0}{\lambda_0+s}. \tag{5.1.4}$$

Finally, from (5.1.1) $\sigma_{0n}(s) = \sigma_0^+(s) \dots \sigma_{n-1}^+(s)$, or

$$\sigma_{0n+1}(s) = \sigma_{0n}(s)\, \sigma_n^+(s). \tag{5.1.5}$$

The following theorem provides insight into the nature of

$s_{0n}(t)$.

Theorem 5.1A. $\sigma_{0N}(s) = \frac{\theta_{N1}}{\theta_{N1}+s} \frac{\theta_{N2}}{\theta_{N2}+s} \cdots \frac{\theta_{NN}}{\theta_{NN}+s}$; $\theta_{Nj} > 0$ where θ_{Nj} are distinct. Before proving the theorem, some comments on the nature of the theorem may be in order.

Remark 5.1B. The theorem shows that the passage time T_{0n} can be thought of as the sum of N independent exponentially distributed r.v.'s. This result is fully analytical, and seems to have no clear probabilisitic interpretation.

Remark 5.1C. A class of probability density functions PF_∞ may be defined for present purposes as the closure of the set of finite convolutions of exponential densities. Thus $s_{0n}(t)$ is PF_∞. (See Karlin, S., *Total Positivity*, Stanford University Press, 1968.) The closure of the class of all PF_∞ functions contains many of the simple p.d.f.'s encountered in statistics (e.g., gamma of integer order, normal).

The proof of the theorem is based on two lemmas,

Lemma 5.1D. Except for singularities, $\sigma_n^+(s)$ is monotonically decreasing for real s.

Proof: Clearly from (5.1.4), $\sigma_0^+(s)$ is monotone decreasing for real s. Supposing $\frac{d}{ds}\sigma_{n-1}^+(s) < 0$, $s \in R$, one finds from (5.1.3) that

$$\frac{d}{ds}\sigma_n^+(s) = \frac{-\lambda_n(1-\mu_n \frac{d}{ds}\sigma_{n-1}(s))}{(s+\lambda_n+\mu_n-\mu_n\sigma_{n-1}^+(s))^2} < 0, \quad (5.1.6)$$

if the denominator is non-zero, by induction.

Lemma 5.1E. $\sigma_n^+(s)$ is a rational function, and has a simple pole between each pair of neighboring poles of $\sigma_{n-1}^+(s)$. All n+1 poles of $\sigma_n^+(s)$ are on the negative halfline.

Proof: Again we proceed by induction. Suppose $\sigma_{n-1}^{+}(s)$ satisfies the statements of the lemma. Then from (5.1.3) $\sigma_n^{+}(s)$ is seen to be rational, and to have poles at the zeros of $s+\lambda_n+\mu_n-\mu_n\,\sigma_{n-1}^{+}(s)$. Using Lemma 5.1D, one finds that between each pair of poles of $\sigma_{n-1}^{+}(s)$, the derivative of $s+\lambda_n+\mu_n-\mu_n\,\sigma_{n-1}^{+}(s)$ is positive. Hence there is exactly one zero in that interval. This zero corresponds to a simple pole of $\sigma_n^{+}(s)$. The other part of the lemma can also be proven inductively. The structure of $\sigma_2^{+}(s)$ is shown in the figure.

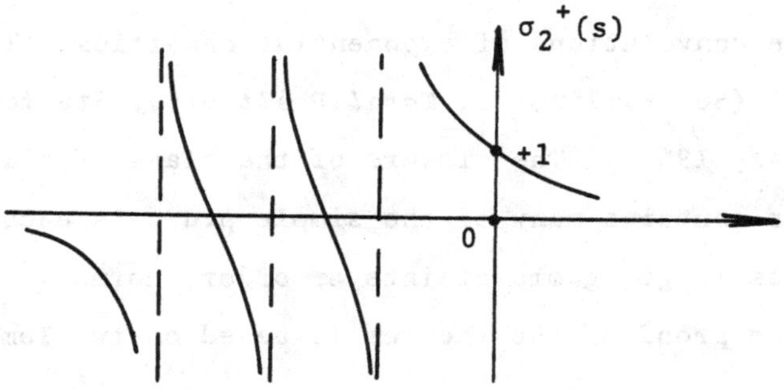

Proof of Theorem 5.1A: The proof is by induction. The case $N = 1$ is trivial. Suppose the theorem holds for N. Then from (5.1.5) with $n = N-1$, it is seen that the zeros of $\sigma_{N-1}^{+}(s)$ are the poles of $\sigma_{0N-1}(s)$, and that the poles of $\sigma_{N-1}^{+}(s)$ are the poles of $\sigma_{0N}(s)$. Now from (5.1.3), the poles of $\sigma_{N-1}^{+}(s)$ are zeros of $\sigma_N^{+}(s)$. Hence the zeros of $\sigma_N^{+}(s)$ coincide with the poles of $\sigma_{0N}(s)$. Therefore in (5.1.5) with $n = N$, it can be seen that $\sigma_{0N+1}(s)$ has the form required. A different proof is given in Keilson, J., [28], *J. Appl. Prob 8* (1971), 391-398.

It would be interesting to know the corresponding structure properties for $s_{nm}(t)$ for $n \neq 0$. In particular is $s_{nm}(t)$ unimodal? This open question has been resolved tentatively in the affirmative.

§5.2. Passage Time Moments for a Birth-Death Process.

Of particular importance are the means and variances of the passage times T_{mn} from state m to state n. Clearly in the notation of §5.1

$$T_{mn} = T_m^+ + T_{m+1}^+ + \dots T_{n-1}^+, \tag{5.2.1}$$

where $T_m^+ = T_{m,m+1}$ has density $s_m^+(\tau)$ and the T_m^+ are independent. Then from (5.1.3) and $-\sigma_m^{+\prime}(0) = E\, T_m^+ = \overline{T}_m^+$, we have

$$\overline{T}_n^+ = \lambda_n^{-1}(1 + \mu_n \overline{T}_{n-1}^+). \tag{5.2.2}$$

The potential coefficients for the chain (§3.3) are given by

$$\pi_0 = 1; \ \pi_n = \frac{\lambda_o}{\mu_1}\frac{\lambda_1}{\mu_2} \dots \frac{\lambda_{n-1}}{\mu_n}, \tag{5.2.3}$$

with $\lambda_n \pi_n = \mu_{n+1} \pi_{n+1}$. If (5.2.2) is multiplied by π_n, we have $(\lambda_n \pi_n \overline{T}_n^+) - (\lambda_{n-1} \pi_{n-1} \overline{T}_{n-1}^+) = \pi_n$. Summation then gives, since $\overline{T}_0^+ = \lambda_0^{-1}$,

$$\overline{T}_n^+ = \frac{1}{\lambda_n \pi_n} \sum_0^n \pi_j .$$

It follows that

$$\overline{T}_{0n} = \sum_{m=0}^{n-1} \frac{1}{\lambda_m \pi_m} \sum_0^m \pi_j . \tag{5.2.4}$$

The reader will verify that $\theta_n^+ = \text{var}\,[T_n^+]$ may be calculated in the same way via differentiation of (5.1.3) twice. One then obtains

$$\theta_n^+ = \frac{\mu_n}{\lambda_n}\theta_{n-1}^+ + \frac{\mu_n}{\lambda_n}(\overline{T}_{n-1}^+)^2 + (\overline{T}_n^+)^2, \tag{5.2.5}$$

and the variance of T_{mn} may be calculated from it. Further discussion may be found in [24].

A similar type result may be exhibited for ruin probabilities defined in the following way. Let the process $N(t)$ commence at a state m with $0 < m < L+1$, and let R_m be the probability that the state $n = L+1$ (ruin) will be reached before the state $n = 0$. Then, since $\lambda_n(\lambda_n+\mu_n)^{-1}$ and $\mu_n(\lambda_n+\mu_n)^{-1}$ are probabilities of a step to the right and left, respectively, from state n, we have

$$\begin{aligned} (\lambda_1+\mu_1)R_1 &= \lambda_1 R_2 \\ (\lambda_m+\mu_m)R_m &= \lambda_m R_{m+1} + \mu_m R_{m-1}, \quad 2 \le m \le L, \end{aligned} \tag{5.2.6}$$

when one adopts the convention that $R_{L+1} = 1$. If we multiply (5.2.6) by π_m and again make use of $\lambda_{m-1}\pi_{m-1} = \mu_m\pi_m$, we may rewrite (5.2.6) as

$$\begin{aligned} 0 &= \lambda_1\pi_1(R_2-R_1)-\lambda_0\pi_0 R_1 \\ 0 &= \lambda_m\pi_m(R_{m+1}-R_m)-\lambda_{m-1}\pi_{m-1}(R_m-R_{m-1}), \quad 2 \le m \le L. \end{aligned} \tag{5.2.7}$$

From this it follows that, for $0 < m < L+1$, $\lambda_m\pi_m(R_{m+1}-R_m) = \lambda_0 R_1$, and summation gives

$$\frac{R_m}{R_1} = 1+\lambda_0 \sum_1^{m-1} (\lambda_j\pi_j)^{-1}, \quad 2 \le m \le L+1. \tag{5.2.8}$$

Since $R_{L+1} = 1$, we have $R_1 = \{1+\lambda_0 \sum_1^{L} (\lambda_j\pi_j)^{-1}\}^{-1}$ and

$$R_m = \frac{1+\lambda_0 \sum_1^{m-1} (\lambda_j\pi_j)^{-1}}{1+\lambda_0 \sum_1^{L} (\lambda_j\pi_j)^{-1}}, \quad 2 \le m \le L+1. \tag{5.2.9}$$

With the help of (5.2.3) this may be written finally as

$$R_m = \frac{\prod_1^L \rho_j + \prod_2^L \rho_j \ \ldots\ldots + \prod_m^L \rho_j}{\prod_1^L \rho_j + \prod_2^L \rho_j \ \ldots\ldots\ldots\ldots + \rho_L + 1}, \quad 1 \leq m \leq L. \quad (5.2.10)$$

where $\rho_j = \lambda_j/\mu_j$.

§5.3. PF_∞, Complete Monotonicity, Log-Concavity and Log-Convexity.

The descriptive distributions for chains in continuous time often take one of several simple forms whose properties are of corresponding interest.

Definition 5.3A. A p.d.f. $f(x)$ is a *Polya frequency function* of infinite order (write $f \in PF_\infty$) if, possibly after translation, $f(x)$ is (the limit of a sequence of densities, each of which is) a convolution of a finite number of exponential densities.

See Karlin, op. cit., for alternative definitions. Examples are the normal density, and the passage time densities $s_{0n}(\tau)$ in a birth-death process. See Theorem 5.1A.

Definition 5.3B. A p.d.f. $f(x)$ on $\mathscr{R}$ is *unimodal* (write $f \in U$) if there exists a number $m \in \mathscr{R}$ with $f(x) \leq f(y)$, for all x,y with $x \leq y \leq m$ or $m \leq y \leq x$.

Definition 5.3C. A p.d.f. $f(x)$ on R is *strongly unimodal* (write $f \in SU$) if $f*g \in U$ for all $g \in U$. Less formally a p.d.f. is strongly unimodal, if it is unimodal and if its convolution with any unimodal p.d.f. is still unimodal.

Easily seen is $f, g \in SU \Rightarrow f*g \in SU$; and $f \in SU \Rightarrow f \in U$. (Take $g_n = \mathscr{N}(0,\frac{1}{n})$. Then $f*g_n \in U$ all n, and hence

$\lim (f*g_n) = f \in U)$.

The following deep, unintuitive and surprising theorem is due to Ibragimov [19]. No proof will be given.

Theorem 5.3D. $f \in SU \longleftrightarrow \log f(\tau)$ is concave on $\{\tau | f(\tau) > 0\}$ and there are no gaps in the interval of support. This is a theorem in $\mathscr{R}^1$ only.

It follows from Theorem 5.3D that since every exponential density is log-concave in the sense of the theorem, it is strongly unimodal. From Theorem 5.1A, we have at once

Theorem 5.3E. For any birth-death process for which $\lambda_j > 0$, $j \geq 0$; $\mu_j > 0$, $j > 0$; $\mu_0 = 0$, the passage time density from state 0 to any state n is PF_∞, hence strongly unimodal and hence log-concave.

A more formal and more convenient definition of log-concavity and log-convexity is the following.

Definition 5.3F. A non-negative density $f(\underline{x})$ on $\mathscr{R}^n$ or a convex subset A thereof is said to be *log-concave* on A if

$$f(\lambda \underline{x} + (1-\lambda)\underline{y}) \geq f^{\lambda}(\underline{x}) f^{1-\lambda}(\underline{y}) \quad 0 \leq \lambda \leq 1 \quad x,y \in A. \tag{5.3.1}$$

Such a function is said to be *log-convex* on A if the reverse inequality holds.

Remark 5.3G.

a) For log-concave p.d.f.'s in $\mathscr{R}^1$, $A = \mathscr{R}^1$ is the set of interest. For log-convex p.d.f.'s, $A = (0,\infty)$ is the set of interest.

b) Log-concavity in $\mathscr{R}^1$ is preserved under convolution. (Because strong unimodality is.)

c) Since $\lambda e^{-\lambda x} U(x)$, where $U(x) = 1$ for $x \geq 0$,

$u(x) = 0$ for $x < 0$, is a log-concave p.d.f., one concludes from (b) and Def. 5.3A that $PF_\infty \subset SU$.

d) $SU = PF_2$ (cf. Karlin [21] for the meaning of PF_2 in the context of total positivity.)

e) Log-concave and log-convex functions are continuous in the interior of their domains. This follows from the corresponding property for convex functions. A good discussion of convexity and related inequalities may be found in Mitrinovic [53].

f) Log-concave functions on $\mathscr{R}^1$ have at most exponential tails, i.e.,

<u>Proposition 5.3H</u>. If $f \in SU$, then

$$f(\tau) = o(e^{-\mu\tau}) \quad \text{for } \tau \to \infty, \text{ for some } \mu > 0. \tag{5.3.2}$$

<u>Proof</u>: $f \in U$. Therefore $f(\tau)$ is nonincreasing for $\tau > m$. For nontriviality, we suppose $f(x) > 0$ for all $x > m$. Then from (5.3.1) with $\lambda = \frac{1}{2}$, $x = n+2$, $y = n$, one finds that $f(n+1)/f(n)$ is nonincreasing and hence tends to a nonnegative limit r, say. The case $r \geq 1$ is dismissed from the integrability of $f(x)$. It is then found that $e^{-\mu} = \frac{1}{2}(r+1)$ satisfies (5.3.2). □

g) In particular, the family PF_∞ which is in SU obeys (5.3.2). Its tails are either asymptotically exponentials as for the Gamma densities, or satisfy (5.3.2) for every $\mu > 0$ as for the normal density.

h) By virtue of Prop. 5.3H, every element of SU has all moments.

i) The strongly unimodal densities have wide prevalence. In particular, most of the classical p.d.f.'s of statistics having all moments are log-concave. The Gamma densities $x^r e^{-x}/\Gamma(r+1)$ with $-1 < r < 0$ are an exception. These are log-convex!

§5.4. Complete Monotonicity and Log-Convexity.

A second family of p.d.f.'s of great importance to the description of Markov chain behavior is that of the completely monotone densities (see Feller II [15], and Widder [66]).

Definition 5.4A. A p.d.f. $f(x)$ on $[0,\infty)$ is *completely monotone* (write $f \in CM$) if all derivatives of f exist and $(-1)^n f^{(n)}(x) \geq 0 \quad n = 1,2,\ldots$.

For completely monotone functions one has Bernstein's theorem, stated here for densities only. (See Feller II.)

Theorem 5.4B. $f \in CM \Longleftrightarrow f(x) = \int_0^\infty ye^{-yx}\, dG(y)$ for some distribution function G, i.e., f is a mixture of exponential densities.

The following basic theorem is due to Artin, E. (vintage 1920). The proof is based on the Holder inequality [53]. We follow the proof of Kingman [47].

Theorem 5.4C. Mixtures of log-convex functions are log-convex.

Proof: Let $\alpha,\beta \geq 0$, $\alpha+\beta = 1$. Let $f(\underline{x})$, $g(\underline{x})$ be log-convex on set A, a convex subset of $\mathcal{R}^n$. Let $h = f+g$. Then

$$\begin{aligned} h(\alpha\underline{x}+\beta\underline{y}) &= f(\alpha\underline{x}+\beta\underline{y}) + g(\alpha\underline{x}+\beta\underline{y}) \\ &\leq f^\alpha(\underline{x})\, f^\beta(\underline{y}) + g^\alpha(\underline{x})\, g^\beta(\underline{y}) \\ &\qquad\qquad \text{c.f. } (5.3.1) \\ &\leq (f(\underline{x}) + g(\underline{x}))^\alpha (f(\underline{y}) + g(\underline{y}))^\beta \quad \text{(Holder)} \\ &= h^\alpha(\underline{x})\, h^\beta(\underline{y}). \end{aligned}$$

It follows that if $f = p_1 f_1 + p_2 f_2$, $p_1, p_2 \geq 0$, $p_1 + p_2 = 1$; and f_1, f_2 are log-convex then f is log-convex. The result extends easily to general mixtures of log-convex functions.

A more special result following trivially from Definition 5.4A is

Prop. 5.4D. Mixtures of completely monotone densities are completely monotone.

The following points are of interest.

Remarks 5.4E.

a) All completely monotone densities are log-convex (from Theorems 5.4B and 5.4C). Complete monotonicity, however, requires differentiability to all orders, and log-convex functions need not be CM. Indeed a function may be log-convex on the positive reals, monotone decreasing, integrable and have derivatives of all order at all $x > 0$, and still not be completely monotone. Examples and discussion may be found in [38].

The intersection of the families CM and SU is the family of purely exponential densities on the positive reals, since only linear functions are both concave and convex.

§5.5. Complete Monotonicity in Time-Reversible Processes.

We have seen earlier (§3.5B) that the passage time densities $s_{n,n+1}(\tau)$ and $s_{n,n-1}(\tau)$ for reaching adjacent states in the birth-death process are completely monotone. It will be seen subsequently that complete monotonicity plays a strong role in the description of the time-dependent behavior of all reversible processes and is a property of their "sojourn time" densities and "exit time" densities.

Complete monotonicity also arises when no time-reversibility is visible. Its prevalence in a somewhat general setting is described in Keilson, J. [33].

§5.6. Some Useful Inequalities for the Families CM and PF_∞.

In the following two theorems, let f be the density of a random variable X, concentrated on $[0,\infty)$.

The first theorem is of particular interest in the context of certain limit theorems. It will be seen subsequently that $(\frac{\sigma^2}{\mu^2} - 1)$ is a distance to pure exponentiality in a metric space sense in the family of completely monotone densities. It is therefore useful as a measure of the exponentiality of such a function.

Theorem 5.6A. $f \in CM \Rightarrow (\frac{\sigma^2}{\mu^2})_X \geq 1.$

Equality occurs if X is exponential.

Proof: From Theorem 5.4B $X = YW$, where Y had density $e^{-x}U(x)$, $W \geq 0$, and Y,W are independent. Then $EX = EY\cdot EW = EW$, $EX^2 = EY^2\cdot EW^2 = 2EW^2$, whence

$$\frac{EX^2-(EX)^2}{(EX)^2} = \frac{2EW^2-(EW)^2}{(EW)^2} = \frac{2\,\sigma_W^2}{\mu_W^2} + 1 \geq 1.$$

Equality occurs if $\sigma_W^2 = 0$, in which case with prob 1 $W = \lambda^{-1}$, say. Then $X = \lambda^{-1}Y$, so that $P\{X > x\} = P\{Y > \lambda x\} = e^{-\lambda x}$. Thus X is exponentially distributed. The converse is trivial.

An analogue of Theorem 5.6A for the class PF_∞ is:

Theorem 5.6B. $f_X \in PF_\infty$, with all support on the positive reals $(\frac{\sigma^2}{\mu^2})_X \leq 1$. Equality occurs if X is exponential.

__Proof__: It can be shown that the property $\frac{\sigma^2}{\mu^2} \leq 1$ is valid for any log-concave p.d.f. with support on the positive reals. The theorem follows from the log-concavity of PF_∞. A simple proof for the case where the density f is the sum of a finite number of independent exponentials E_i is the following. Suppose $X = E_1 + E_2 + \ldots + E_n$, where E_i are independent and exponentially distributed with mean μ_i. Then $\sigma_X^2 = \sum_1^n \mu_i^2 \leq (\sum_1^n \mu_i)^2 = \mu_X^2$, so that the first part is proven for this case. Since $\Sigma_1^n \mu_i^2 = (\sum_1^n \mu_i)^2$ if and only if only one term in $\Sigma \mu_i$ is non zero, the second statement is also proven for the finite case. For the general case, the reader is referred to Karlin, S., Proschan, F., and Barlow, R. E., "Moment inequalities of Polya frequency functions," *Pacific J. Math. 11* (1961), 1023-1033, and Keilson, J., "A threshold for log-concavity for probability generating functions and associated moment inequalities," *Ann. Math. Stat. 43* (1972), 1702-1708.

A more complete set of inequalities for the moments of densities which are log-concave or log-convex is available [29]. The results are stated next without proof.

__Theorem 5.6C__. Let the p.d.f. $f(x)$ have all support on the positive reals and be log-concave there. Then $f(x)$ has all moments and

$$\left\{\frac{\mu_{k+1}}{(k+1)!}\right\}^{\frac{1}{k+1}} \leq \left\{\frac{\mu_k}{k!}\right\}^{1/k} \quad k = 1,2,\ldots .$$

If $f(x)$ is log-convex there, then the inequalities are reversed. (μ_k may become infinite beyond some point in the sequence.)

An example of a log-convex p.d.f. whose moment sequence truncates is $K_r(1+x)^{-r}$ where r is any real number greater than one.

§5.7. Log-Concavity and Strong Unimodality for Lattice Distributions.

The previous results for unimodality, strong unimodality and log-concavity are easily extended to discrete distributions, with comparable results [34].

Definition 5.7A. A distribution (p_n) with all support on the lattice of integers will be said to be *unimodal* if there exists at least one integer M such that

$$p_n \geq p_{n-1} \quad \text{all} \quad n \leq M$$

$$p_{n+1} \leq p_n \quad \text{all} \quad n \geq M.$$

In this case we write $(p_n) \in U$.

Clearly lattice unimodality implies a single lattice interval of support, i.e., no gaps.

Definition 5.7B. A discrete distribution (h_n) is *strongly unimodal* if the convolution of (h_n) with any unimodal (p_n) is unimodal. In this case we write $(h_n) \in SU$.

Since a distribution with all mass at zero is unimodal, $SU \subset U$. The definition easily implies that SU is closed under convolution.

The following important theorem is the analogue to Theorem 5.3D. It will not be proven here [34].

Theorem 5.7C. A necessary and sufficient condition that (p_n) be strongly unimodal is that (p_n) be log-concave, i.e., that

$$p_n^2 \geq p_{n+1}\, p_{n-1} \quad \text{all} \quad n. \tag{5.7.1}$$

One has "log-concavity" in the sense that $\Delta^{(2)} \log p_n \leq 0$ in the language of finite differences.

It is easily verified that the above condition implies the analogue of (5.3.1), for all integers n and positive integers m_1, and m_2.

$$p_n \geq p_{n+m_1}^{(m_2/m_1+m_2)}\, p_{n-m_2}^{(m_1/m_1+m_2)}.$$

An analogue to (5.3.2) follows from the same proof there. That is, if $(p_n) \in SU$ then

$$p_n = o(e^{-\mu n}), \text{ as } n \to \infty \text{ for some } \mu > 0. \tag{5.7.2}$$

It follows that all moments of the distribution (p_n) exist if it is strongly unimodal.

5.7D. Prevalence of strong unimodality.

a) The Bernoulli distribution trivially satisfies (5.7.1) and is strongly unimodal.

b) A binomial distribution is equivalent to a finite convolution of Bernoulli distributions and hence is strongly unimodal. Similarly any distribution which is the convolution of independent Bernoulli distributions with different parameters is strongly unimodal.

c) The Poisson distribution also satisfies (5.7.1) as the reader will verify or may be thought of as the limit of binomial distributions. (Clearly log-concavity is preserved under limits.)

d) The geometric distribution satisfies (5.7.1) exactly and is the discrete analogue of the exponential distribution.

e) Negative binomial distributions may be shown from simple algebra to satisfy (5.7.1).

f) Tail sums of strongly unimodal distributions are strongly unimodal. (See 5.8A.)

Two examples illustrating the use of these log-concavity ideas for lattices will now be given.

Example 5.7E. Consider a K out of N reliability systems whose components $C_1, C_2, \ldots C_N$ are independent with ergodic probabilities $e_j = P[C_j$ is working]. Let X_j be the performance indicator for C_j, i.e., $X_j = 1$ when C_j works, $X_j = 0$ otherwise. Then the number of working components at ergodicity is $Y = \sum_1^N X_j$ and the ergodic probability $f_n = P[Y = n]$ is the convolution of the ergodic probabilities of the component performance indicators, each of which is distributed on $\{0,1\}$, i.e., is Bernoulli. Since Bernoulli distributions are log-concave (5.7Da), the distribution (f_n) is strongly unimodal, i.e., unimodal. The moment inequalities of the next section are then valid.

Example 5.7F. Consider an ergodic birth-death process with $\lambda_n/\mu_{n+1} \downarrow n$. We then have for the ergodic probabilities e_n, from detailed balance,

$$\frac{e_{n+1}}{e_n} = \frac{\lambda_n}{\mu_{n+1}} \quad \downarrow n.$$

A special example is the following variant on the homogeneous Poisson queue for the number of customers in the system. The arrival stream at rate λ may be subject to "balking", i.e., reluctance of join the queue when the line is long. Then

λ_n decreases as n increases. There may also be "reneging", i.e., departures from the line when it gets long, so that μ_n increases with n. Consequently λ_n/μ_{n+1} decreases with n. From $e_{n+1}/e_n \leq e_n/e_{n-1}$ we have $e_n^2 \geq e_{n+1} e_{n-1}$, i.e., strong unimodality.

§5.8. Preservation of Log-Concavity and Log-Convexity under Tail Summation and Integration.

Log-concavity of (p_n) with $p_n^2 \geq p_{n+1} p_{n-1}$, has the equivalent characterization p_{n+1}/p_n decreases as n increases. This may be used to show the following.

Prop. 5.8A. If (p_n) is log-concave, then the tail sum masses (P_n) where $P_n = \sum_n^\infty p_r$ are also log-concave. If (p_n) is log-convex on $\mathcal{N} = \{n = 0,1,2,\ldots\}$, i.e., if $p_n^2 \leq p_{n+1} p_{n-1}$, $n \geq 1$, then so also are (P_n), i.e., $P_n^2 \leq P_{n+1} P_{n-1}$, $n \geq 1$.

The proof of the proposition follows from the identity

$$P_N^2 = P_{N+1} P_{N-1} + \sum_{j=N}^\infty (p_N p_j - p_{N-1} p_{j+1}), \qquad (5.8.1)$$

which the reader will verify. In the log-concave case one has $p_{j+1}/p_j \leq p_N/p_{N-1}$ when $j \geq N$. Hence log-concavity of (p_n) implies that the summands on the right of (5.8.1) are positive so that $P_N^2 \geq P_{N+1} P_{N-1}$ for all N[†]. In the log-convex case one does not have or require the inequality $P_0^2 \leq P_{+1} P_{-1}$ and the same argument is available.

It is easy to see heuristically that if f_X is a log-concave (log-convex) p.d.f. on the continuum, then

$$\overline{F}_X(x) = \int_x^\infty f(y)dy = P[X > x],$$

[†]A more intuitive proof is that the sequence $(\sum_0^N p_m) = (P_0, P_1, P_2, \ldots\ldots) * (1,1,1,\ldots\ldots)$ is log-concave since the sequence $(1,1,1,\ldots\ldots)$ is log-concave. Similarly $(\sum_N^K p_m)$ is log-concave in N for K fixed and one may let $K \to \infty$.

the so-called "*survival function*" of the distribution of X, is also log-concave (log-convex). One need only visualize the density as the limit in measure of a sequence of lattice distributions when the grain of the lattice goes to zero and apply the theorem above. A careful demonstration of the preservation for the continuum use can be based on an integral inequality similar in structure to that of the identity used to prove Theorem 5.8A [31]. Formally we have

Prop. 5.8B. If $f(x)$ is log-concave then $\overline{F}(x)$, its survival function, is log-concave. If $f(x)$ is log-convex, then $\overline{F}(x)$ is log-convex.

§5.9. Relation of CM and PF$_\infty$ to IFR and DFR Classes in Reliability.

Let $a(x)$ be a p.d.f. on the positive reals and suppose for simplicity that $a(x)$ is positive and differentiable for all x. Let $\overline{A}(x)$ be the survival function and let

$$\eta(x) = -\frac{d}{dx}\log \overline{A}(x) = \frac{a(x)}{\overline{A}(x)}. \tag{5.9.1}$$

Then

$$\overline{A}(x) = \exp\{-\int_0^x \eta(y)dy\} = e^{-N(x)}. \tag{5.9.2}$$

The entity $\eta(x)$ is called the failure rate or hazard rate of the distribution and the distribution is said to be IFR when $\eta(x)$ is monotone increasing, and DFR when $\eta(x)$ is monotone decreasing. Clearly the log-convexity of $\overline{A}(x)$ is equivalent in this case to the DFR property, and the log-concavity of $\overline{A}(x)$ to the IFR property. In particular, since PF$_\infty$ corresponds to $a(x)$ log-concave (§5.3) and CM corresponds to $a(x)$ log-convex it follows from Theorem 5.8B that

Prop. 5.9A.

$$PF_\infty \subset IFR, \quad \text{and}$$
$$CM \subset DFR.$$

A much more extensive and precise treatment of class interrelationships may be found in [4] and [5].

Chapter 6

Passage Times and Exit Times for More General Chains

§6.00. Introduction.

To assess the time to failure to complex systems, describe the course of an epidemic, quantify the dynamics of the human metabolic system or describe congestion in networks via general Markov chain models, one must partition the state space into two sets, G and B say, and study the transfer between them. Two techniques available are examined. The first employs uniformization to exhibit explicitly the mean time from states of G to the set B and higher moments in terms of the Green potential matrix of the set G. One also finds in the same way ruin probabilities of interest. The notion of ergodic transfer rates or flow rates between states and sets is developed and employed to define various transfer times of interest, specifically the "ergodic exit time" from set G, the "quasi-stationary exit time", and the "ergodic sojourn time" on set G. For time-reversible systems, the simplicity of the theoretical structure present permits one to exhibit a stochastic order between these. The sojourn time is of special interest because its mean may be expressed in terms of the ergodic probabilities and transfer rates between states. The structural property of complete monotonicity and its DFR character plays an important role. A certain element of "jitter", i.e., multiple crossings at the boundary between the good and bad states, jeopardizes the validity of the sojourn time as a system failure time and makes the exit time more desirable for such a role.

§6.1. Passage Time Densities to a Set of States.

The study of the transient behavior of a Markov chain $N(t)$ in continuous time often requires the solution of the following problem. The chain starts in some particular state, $N = 0$ say, and one is interested in the distribution of the first time T_{0B} at which a specified subset B of the state space is reached. For the finite chain of interest to us, this "first passage time" will have an absolutely continuous density.

When the subset B consists of a single state b, the passage time density $s_{0b}(\tau)$ obeys the equation

$$p_{0b}(t) = s_{0b}(t) * p_{bb}(t), \qquad (6.1.1)$$

where $p_{mn}(t)$ is the transition probability, and the asterisk denotes convolution. For the Laplace transforms, we then have $\sigma_{0b}(s) = \pi_{0b}(s)/\pi_{bb}(s)$ and only inversion is needed, but the inversion may be clumsy or intractable. When the transition probabilities are known explicitly a trick is useful. One may write

$$\sigma_{0b}(s) = \frac{s\,\pi_{0b}(s)}{1-\{1-s\,\pi_{bb}(s)\}} = \frac{L\{\frac{d}{dt}\,p_{0b}(t)\}}{1-L\{\frac{-d}{dt}\,p_{bb}(t)\}}\,. \qquad (6.1.2)$$

The value of the denominator at $s = 0$ is $1-(1-e_b) = e_b > 0$. Differentiation at $s = 0$, always permissible for finite chains, then gives for $E\ T_{0b} = -\sigma'_{0b}(0)$,

$$E\ T_{0B} = e_b^{-1}\{\int_0^\infty (e_b - p_{0b}(t))dt + \int_0^\infty (p_{bb}(t) - e_b)dt\}$$

$$= e_b^{-1}\{\int_0^\infty (p_{bb}(t) - p_{0b}(t))dt\}. \qquad (6.1.3)$$

Example 6.1A. Suppose a system consists of K independent Markov components with different failure time and repair time parameters, and suppose that all components must fail for the system to fail. If the system starts with all components working, the failure time is the passage time from the state with all components up to the state with all components down. Because of the independence, the ingredients needed for the explicit evaluation of (6.1.3) are at hand. For this example the entities in curly brackets on the right hand side of (6.1.2) are positive density functions. If we write (6.1.2) as

$$\sigma_{0b}(s) = \alpha(s)[1-\beta(s)]^{-1} = \sum_0^\infty \alpha(s)\ \beta^k(s), \qquad (6.1.4)$$

we then have, since $\beta(0) = 1-e_b < 1$,

$$s_{0b}(\tau) = \sum_0^\infty a(\tau) * b^{(k)}(\tau). \qquad (6.1.5)$$

This formal solution converges slowly when e_b is small. An explicit solution may be obtained from (6.1.4) by the method of residues, since (cf. §5.1) the zeros of $1-\beta(s)$ lie on the negative real s-axis. These zeros and the residues needed may be obtained from $p_{0b}(t)$ numerically with computer assistance.

The passage time density and moments from a state to a

set B with more than one state is more difficult. To find such a density it is necessary to introduce a vector p.d.f. $\underline{s}_{0B}(\tau)$ in place of the scalar p.d.f. one uses for singleton B set. The vector density $\underline{s}_{0B}(\tau)$ has components $s_{0m}(\tau)$ for each of the states m of the set B, where $s_{0m}(\tau)$ is the joint probability density that B is reached at time τ after departure from 0 and that arrival to the set occurs at state m. For this vector density function, one has

$$\sum_B \int_0^\infty s_{0m}(\tau)\, d\tau = 1, \tag{6.1.6}$$

and

$$s_{0B}(\tau) = \sum_B s_{0m}(\tau) = \underline{s}_{0B}^T(\tau)\, \underline{1} \tag{6.1.7}$$

is the scalar p.d.f. of the passage time from state 0 to set B. The mean passage time is

$$E\, T_{0B} = -[\underline{\sigma}'(0)]^T\, \underline{1} = \Sigma \int_0^\infty \tau\, s_{0m}(\tau)\, dt. \tag{6.1.8}$$

Of related interest are what may be called *ruin probabilities*. The classical gambler's ruin problem [14] deals with a homogeneous random walk $X_{k+1} = X_k + \xi_k$, with ξ_k i.i.d. random variables having a given distribution on $\{-1,+1\}$. The process starts at $X_0 = n_0$, $1 \le n_0 \le L-1$, and the ruin probability is the probability of hitting the state 0 before hitting L. For the process N(t) of interest here, we define the ruin probabilities

$$R_0^m = P[\text{set B is first reached from state 0 to state m}]. \tag{6.1.9}$$

Clearly

$$R_0^m = \int_0^\infty s_{0m}(\tau) d\tau. \tag{6.1.10}$$

It is natural to try to evaluate the ruin probabilities

R_0^m and mean passage time $E\,T_{0B}$ from some analogue of (6.1.1) and (6.1.2). Such an analogue is given by the set of equations

$$p_{0n}(t) = \sum_B s_{0m}(t) * p_{mn}(t), \quad n \in B. \tag{6.1.11}$$

This comes about from the observation that to be at state n at time t, the set B must be reached for the first time at some t' in $(0,t)$ and some state m, and then be found subsequently at state n at time t. The equation set (6.1.11) has the vector form

$$\underline{p}_{0B}^T(t) = \underline{s}_{0B}^T(t) * \mathfrak{p}_B(t) \tag{6.1.12}$$

where $\mathfrak{p}_B(t)$ denotes the restriction of the transition matrix $\mathfrak{p}(t)$ to the components of set B. Laplace transformation of (6.1.12) leads to the formal solution

$$\underline{\sigma}_{0B}^T(s) = \underline{\pi}_{0B}^T(s)\ [\pi_B(s)]^{-1}, \tag{6.1.13}$$

because of the singularity of the components $\pi_{mn}(s)$ at $s = 0$ and questions of invertibility of the matrix $\pi(s)$, progress via (6.1.13) is blocked.† We will see subsequently that these difficulties may be overcome by taking two alternate routes. The first procedure calculates explicitly the mean passage time and ruin probabilities with the help of the Green potentials of Chapter 4. The second utilizes the compensation method of Chapter 4 to reduce the rank of the problem to the number of states in B. Each procedure has advantages and disadvantages.

---

†Actually the procedure leading to (6.1.3) goes through agreeing with the results later obtained.

§6.2. Mean Passage Times to a Set via the Green Potential.

Let a continuous time, irreducible Markov chain $N(t)$ on a finite state space $\mathcal{N}$ be governed by hazard rates (ν_{mn}). The chain is uniformizable, and is therefore governed by $[\nu, a_\nu]$, as described in §2.1. Suppose $N(t)$ represents the behavior of a system, where the total set of states $\mathcal{N}$ is partitioned into a set of "good" states G where the system is working properly, and a "bad" set B of states where the system is not working properly. In congestion theory, B might indicate saturation of the system. For such a partition, $\mathcal{N} = G \cup B$, $G \cap B = \phi$. One is interested in the passage time density $s_{mB}(\tau)$ from state $m \in G$ to the set B, and in its moments $E[T_{mB}^k]$, $k = 1,2,\ldots$.

The uniformizability of the chain permits one to find expressions for $E[T_{nB}^k]$, which are suited for machine evaluation. The key argument used in the derivation is the self-consistency relation

$$s_{mB}(\tau) = \nu e^{-\nu t} * \{\sum_{n\in G} a_{\nu;mn}\, s_{nB}(\tau) + \sum_{j\in B} a_{\nu;mj}\, \delta(\tau)\}, \quad (6.2.1)$$

where $\delta(\tau)$ is the generalized density for all support at 0. Equation (6.2.1) can be understood as follows: At $m \in G$ there is an exponential dwell time, followed by a transition (possibly to m itself) to either a state $n \in G$ or a state $j \in B$. If the transition is to $n \in G$, one has to add to the dwell time at m the passage time from n to B, i.e., one convolves the dwell time density with $s_{nB}(\tau)$. If the transition is to $j \in B$, the passage time from m to B is completed, and one adds 0 to the dwell time.

Now, taking Laplace transform in (6.2.1), one obtains

$$\sigma_{mB}(s) = \frac{\nu}{\nu+s} \left[\sum_{n\in G} a_{\nu;mn} \sigma_{nB}(s) + \sum_{j\in B} a_{\nu;mj} \right]. \tag{6.2.2}$$

Expectations are found by differentiation at $s = 0$. When use is made of $\sigma_{nB}(0) = 1$ for all $n \in G$ and $\sum_{n\in B\cup G} a_{\nu;mn} = 1$, one obtains

$$E\, T_{mB} = \frac{1}{\nu} + \sum_{n\in G} a_{\nu;mn}\, E\, T_{nB}. \tag{6.2.3}$$

In matrix notation, writing $\underline{T}_G$ for the vector with components $E\, \underline{T}_{mB}$, $m \in G$, and $\mathbf{a}_{\nu G}$ for the submatrix of $\mathbf{a}_\nu$ on $G \times G$, (6.2.3) becomes $[I_G - \mathbf{a}_{\nu G}]\underline{T}_G = \nu^{-1} \underline{1}_G$. Hence, from §4.1,

$$\underline{T}_G = \frac{1}{\nu} [I - \mathbf{a}_{\nu G}]^{-1} \underline{1}_G = \frac{1}{\nu} \mathbf{g}_G \underline{1}_G, \tag{6.2.4}$$

where $\mathbf{g}_G$ is the Green potential of $\mathbf{a}_{\nu G}$. Explicitly

$$E\, T_{mB} = \frac{1}{\nu} \sum_{n\in G} g_{G;mn} = \frac{1}{\nu} \sum_{k=0}^{\infty} \sum_{n\in G} (\mathbf{a}_{\nu G}^k)_{mn}. \tag{6.2.5}$$

More generally, for higher moments, the following theorem holds.

<u>Theorem 6.2A</u>. Let $N(t)$ be a finite Markov chain on state space $\mathcal{N}$. Let $\mathcal{N} = G+B$ where B is a proper subset of $\mathcal{N}$ and G, B are disjoint. Let T_{mB} be the random passage time from $m \in G$ to the set B. Then for any positive integer k,

$$E[T_{mB}^k] = \frac{k!}{\nu^k} \sum_{n\in G} (\mathbf{g}_G^k)_{mn}. \tag{6.2.6}$$

<u>Proof</u>: Equation (6.2.2) may be rewritten as

$$[(s+\nu) I_G - \nu\, \mathbf{a}_{\nu G}] \underline{\sigma}_G(s) = \nu\, \mathbf{a}_{\nu GB}\, \underline{1}_B. \tag{6.2.7}$$

The vectors needed are $E[\underline{T}_G^{(k)}] = (-1)^k[(\frac{d}{ds})^k \underline{\sigma}_G(s)]_{s=0}$. If (6.2.7) is differentiated k times and Leibniz's rule for the differentiation of products is employed one finds

$$\nu(I_G - a_{\nu G})(-1)^k \underline{T}_G^{(k)} + k(-1)^{k-1} \underline{T}_G^{(k-1)} = \underline{0} \quad (6.2.8)$$

and the theorem follows.

When the set B is rare, i.e., when $\sum_B e_n << 1$, the substochastic matrix $a_{\nu G}$ has a maximal eigenvalue close to 1 and the series expansion $g_G = \sum_0 a_{\nu G}^k$ converges slowly. It may be noted that an explicit expression for $\underline{s}_G(\tau)$ is given by

$$\underline{s}_G(\tau) = -\frac{d}{dt}[e^{-\nu\tau[I-a_{\nu G}]} \underline{1}_G]. \quad (6.2.9)$$

This may be seen probabilisitcally or derived in the following manner. Equation (6.2.7) may be rewritten as

$$[I_G - \frac{\nu}{\nu+s} a_G] \underline{\sigma}_G(s) = \frac{\nu}{\nu+s} a_{GB} \underline{1}_B.$$

From this we obtain

$$\sigma_G(s) = \sum_{k=0}^{\infty} (\frac{\nu}{\nu+s})^{k+1} a_G^k a_{GB} \underline{1}_B.$$

Inversion of the transform then gives

$$\underline{s}_G(t) = \nu \sum_{k=0}^{\infty} \frac{(\nu t)^k}{k!} e^{-\nu t} a_G^k a_{GB} \underline{1}_B.$$

When use is made of

$$a_{GB} \underline{1}_B = \underline{1}_G - a_{GG} \underline{1}_G = (I_G - a_{GG})\underline{1}_G$$

(6.2.9) is obtained.

§6.3. Ruin Probabilities via the Green Potential.

The ruin probabilities defined by (6.1.9) have a more general setting. We give two motivating examples.

Example 6.3A. Consider the following problem in reliability theory. A system consists of K independent Markov components, governed by failure rate λ_k, and repair rate μ_k for component k. Failure is defined by arrival at any state of a specified bad set B, a proper subset of the state space $\mathcal{N} = G+B$. The bad set is always entered at one of a number of "boundary" states $B^* \subset B$. To estimate the importance of the candidate entry states, one wishes to evaluate the ruin probabilities R_m for $m \in B^*$ defined by (6.1.9) when the system starts with all components working, designated the perfect state 0.

Example 6.3B. A population of individuals is subject to an epidemic [3]. At time t there are $S(t)$ susceptibles and $I(t)$ infectives. In a small time interval dt there is probability $c_1 S(t)I(t)dt + o(t)$ that a susceptible becomes an infective, and probability $c_2 I(t)dt + o(t)$ that an infective leaves the system through death or recovery. The epidemic is governed by a continuous time Markov chain with states (s,i) $s+i \leq N_0$. $S(t) + I(t)$ is monotone decreasing in t. Interest in the process ends at the set $S = \{(s,0)\ s = 1,\ldots,N_0\}$ (no infectives), or at the set $I = \{(0,i)\ i = 1,\ldots,N_0\}$ (no susceptibles). One wishes to know the time for the epidemic to run its course, and probabilities that the epidemic will finish leaving given numbers of the population uninfected. In our notation $\mathcal{N} = G+B$, $B = S+I$.

In these problems, one starts at $m \in G$, and one of the states B must be reached. The set of states B is itself partitioned as $B = A+\bar{A}$ and ruin is defined as reaching A before $\bar{A}$. The ruin probabilities R_m^A are defined by

$$R_m^A = P[A \text{ is reached before } \bar{A} \mid \text{start at } m]. \tag{6.3.1}$$

Then, as explained below,

$$R_m^A = \sum_G a_{mn} R_n^A + \sum_A a_{mn}; \quad m \in G. \tag{6.3.2}$$

Here $\mathbf{a}$ is the matrix of transition probabilities, if one deals with a discontinuous time chain, or $\mathbf{a}$ is the $\mathbf{a}_\nu$ matrix if the chain is continuous time and uniformizable. (6.3.2) is explained as follows: from state m, one either goes to some state n in G (first sum) or one goes immediately to A (second sum), or one goes to some state $\bar{A}$ (contributing 0 to R_m^A).

In vector notation (6.3.2) reads

$$\underline{R}^A = \mathbf{a}_G \underline{R}^A + \mathbf{a}_{GA} \underline{1}_A, \tag{6.3.3}$$

so that, as before,

$$\underline{R}^A = (\mathbf{I}_G - \mathbf{a}_G)^{-1} \mathbf{a}_{GA} \underline{1}_A = \mathbf{g}_G \mathbf{a}_{GA} \underline{1}_A, \tag{6.3.4}$$

and the non-singularity of $\mathbf{I}_G - \mathbf{a}_G$ is again assured by the strictly substochastic character of $\mathbf{a}_G$. The form (6.3.4) is machine computable, but convergence of $\mathbf{g}_G = \sum_0^\infty \mathbf{a}_G^k$ is slow when the maximal eigenvalue of $\mathbf{a}_G$ is close to one.

For the ruin probabilities of (6.1.9) A is the singleton set $\{m\}$ in B, and (6.3.4) gives

$$R_0^m = \sum_{k=0}^{\infty} \sum_{n \in G} (\mathbf{a}_G^k)_{0n} \, a_{nm}. \tag{6.3.5}$$

§6.4. Ergodic Flow Rates in a Chain.

Detailed balance in a stationary ergodic chain $N(t)$ was defined in §2.4 by $e_m\nu_{mn} = e_n\nu_{nm}$ for all m,n in the state space $\mathscr{N}$. There it was stated without proof that $e_m\nu_{mn}$ was the "probabilistic flow rate" from m to n for the stationary process. The ergodic flow rates have great value as tools for quantifying the time-dependent behavior of such chains, as we will see. A demonstration that these "flow rates" have sharper meaning as asymptotic renewal densities for the transitions from m to n may help justify the term "flow rates".

Let (m,n) be any ordered pair of states of $\mathscr{N}$ for which $\nu_{mn} > 0$. With $N(t)$ we associate an auxiliary process $T(t)$ defined on a state space whose elements are all such ordered pairs. Let $T(t) = (k,\ell)$ when $N(t) = \ell$ and the previous state of $N(t)$ was k. Then it is easy to see that $T(t)$ is itself a Markov chain governed by transition rates

$$\nu_{(m,n)(k,\ell)} = \delta_{nk}\,\nu_{k\ell}, \tag{6.4.1}$$

and that $T(t)$ is ergodic. Consider the epochs $(\tau_{mn}^{(j)})_{j=1}^{\infty}$ at which $N(t)$ has transitions from m to n. Then these epochs are the return times of $T(t)$ to the state (m,n), and the intervals between them are independent and identically distributed. The number of such epochs in $(0,t)$ is then a renewal process whose asymptotic renewal density $h_{mn} = \mu_{(m,n)}^{-1}$ where $\mu_{(m,n)}$ is the mean time of return to state (m,n). Then $\mu_{(m,n)} = E\,D_{mn} + E\,B_{mn}$ where D_{mn} is the dwell time in (m,n) and B_{mn} is the time outside (m,n) until return, i.e.,

$$h_{mn} = \frac{1}{E[D_{mn}] + E[B_{mn}]}. \tag{6.4.2}$$

The ergodic probabilities $e^*_{(m,n)}$ of $T(t)$ satisfy

$$e^*_{(m,n)} = \frac{E[D_{mn}]}{E[D_{mn}] + E[B_{mn}]}. \tag{6.4.3}$$

From (6.4.1) it is seen that the exponentially distributed dwell time D_{mn} has expectation $\nu_n^{-1} = (\Sigma\ \nu_{n\ell})^{-1}$, so that, combining (6.4.2) with (6.4.3) one obtains

$$h_{mn} = e^*_{(m,n)}\ \nu_n. \tag{6.4.4}$$

It remains to find $e^*_{(m,n)}$. This is most conveniently done by considering the following differential equation

$$\frac{d}{dt} P\{T(t) = (m,n)\} = -\nu^*_{(m,n)} P\{T(t) = (m,n)\} + \sum_{(k,\ell)} \nu^*_{(k,\ell)(m,n)} P\{T(t) = (k,\ell)\}. \tag{6.4.5}$$

Here $\nu^*_{(m,n)} = \sum_{(k,\ell)} \nu^*_{(m,n)(k,\ell)}$, and from (6.4.1) it is seen that $\nu^*_{(m,n)} = \nu_n$. Similarly, the second term in the right side of (6.4.5) equals

$$\Sigma_k\ \nu_{mn}\ P\{T(t) = (k,m)\} = \nu_{mn}\ P\{N(t) = m\}. \tag{6.4.6}$$

Setting the left side of (6.4.5) to zero, because of stationarity, gives with (6.4.6)

$$0 = -\nu_n\ e^*_{(m,n)} + \nu_{mn}\ e_m, \tag{6.4.7}$$

so that $e^*_{(m,n)} = \dfrac{e_m\ \nu_{mn}}{\nu_n}$.

Hence, from (6.4.4) $h_{mn} = e_m\ \nu_{mn}$, as one expected from probabilistic intuition.

§6.5. Ergodic Exit Times, Ergodic Sojourn Times, and Quasi-Stationary Exit Times.

Consider an ergodic Markov chain $N(t)$ on state space $\mathcal{N}$ and a partition $\mathcal{N} = G+B$ into good and bad states. We have in mind complex repairable systems such as in Example 6.3A. A second example of a similar character is the following.

Example 6.5A. A communcation system consists of a network with nodes $N_1, N_2, \ldots N_k$ which are cities, and edges $E_1, E_2, \ldots E_k$ which are telephone links. The links are either working or not working and all links are independent. Each link is Markov and governed by two parameters, a failure rate and repair rate. The system is working when all cities are connected by some path, i.e., some set of working links.

These examples are modeled by time-reversible processes, but the results of this section are valid for any ergodic chain, and extend easily to more general systems [30].

In such a general setting one wants to define a sensible "failure time" distribution. A passage time density $s_{m_0B}(\tau)$ could be of interest if there is a particular state m_0 of significance. For Example 6.5A or more general complex repairable systems, a natural state of interest is the state where all components of the system are in perfect condition - in our example all links would be working. But even in such systems the perfect state may be unnatural in that the system is highly redundant and is constantly under repair in its working condition, and the perfect state is rarely visited. One needs to consider passage times to the bad set B which are natural mixtures of the passage times from the good states of G.

Two candidates of interest are the *ergodic exit time from G* and the *ergodic sojourn time on G*. For the ergodic exit time one specifies that at $t = 0$, the system has been running since time immemorial and has been unobserved since its inception but is known to be working, i.e., to be in the good set G. This is equivalent to specifying the initial distribution

$$p_m(0) = \frac{e_m}{\sum_G e_m} = \frac{e_m}{P(G)}, \quad m \in G, \tag{6.5.1a}$$

$$p_m(0) = 0, \quad m \in B. \tag{6.5.1b}$$

A second candidate of interest is the *ergodic sojourn time*, or ergodic visit time to the set G. Here one visualizes visits to set G alternating with visits to set B in the stationary process, and the states n of G for this exit time are weighted by the relative frequency with which visits to G commence at state m. We have seen in §6.4 that the asymptotic renewal density for transitions from m to n is $e_m \nu_{mn} = h_{mn}$. Clearly for the stationary process,

$$\begin{aligned} h_{Bn} &= \lim_{t\to\infty} \frac{1}{t} \text{ [number of transitions from B to n in } (0,t)] \\ &= \sum_{m\in B} e_m \nu_{mn}, \end{aligned} \tag{6.5.2}$$

i.e., h_{Bn} is the ergodic flow rate from B to $n \in G$. For the ergodic sojourn time, therefore, we specify

$$p_m(0) = \frac{h_{Bm}}{\sum_G h_{Bm}}, \quad m \in G, \tag{6.5.3a}$$

$$p_m(0) = 0, \quad m \in B. \tag{6.5.3b}$$

With these initial weightings, the ergodic exit time density $s_E(\tau)$ and the ergodic sojourn time density $s_V(\tau)$ are given by

$$s_E(\tau) = \frac{\sum_G e_n \, s_{nB}(\tau)}{\sum_G e_n}, \tag{6.5.4}$$

and

$$s_V(\tau) = \frac{\sum_G h_{Bn} \, s_{nB}(\tau)}{\sum_G h_{Bn}}, \tag{6.5.5}$$

i.e.,

$$s_V(\tau) = \frac{\sum_{m\in B} \sum_{n\in G} e_m \nu_{mn} \, s_{nB}(\tau)}{\sum_{m\in B} \sum_{n\in G} e_m \nu_{mn}} \tag{6.5.6}$$

§6.6. The Quasi-Stationary Exit Time. A Limit Theorem.

A third candidate of interest is the "*quasi-stationary exit time*". Here it is known that the system has been working since time immemorial. Of course, the probability of this event is zero, but we have in mind the idealization associated with the quasi-stationary distribution [11,59]. The context is again that of an ergodic chain $N(t)$ on $\mathcal{N} = G+B$ of the previous section. One further restriction is needed. The set G must itself be irreducible in the sense that every state of G may be reached from every other state without passing through B. The probability state vector $\underline{p}_G^*(t)$ at time t, given that the system was in state $n \in G$ at $t = 0$ and that the system did not leave G in $(0,t)$ then is

$$\underline{p}_G^*(t) = \frac{\underline{u}_n^T \, e^{-\nu t[I_G - a_G]}}{\underline{u}_n^T \, e^{-\nu t[I_G - a_G]} \, \underline{1}_G}, \tag{6.6.1}$$

where $\underline{u}_n^T$ is the initial state vector with all support at n.

We employ in (6.6.1) the spectral representation (to be proven soon)

$$e^{-\nu t[I_G-\mathbf{a}_G]} = e^{-\nu t(1-\theta_{\nu G})}\{J_G + \epsilon_G(t)\}, \tag{6.6.2}$$

where J_G is the principal idempotent spectal dyad of $\mathbf{a}_G$, $\theta_{\nu G}$ is the associated Perron-Romanovski-Frobenius eigenvalue of $\mathbf{a}_G$ and $\epsilon_G(t) \to 0$ as $t \to \infty$. One obtains, using $J_G = \frac{\underline{r}\underline{\ell}^T}{\underline{\ell}^T\underline{r}}$, where $\underline{r},\underline{\ell}^T$ are positive,

$$\lim_{t\to\infty} \underline{p}_G^*(t) = \frac{(\underline{u}_n^T\ \underline{r})\underline{\ell}^T}{(\underline{u}_n^T\ \underline{r})\underline{\ell}^T\cdot\underline{1}} = \frac{\underline{\ell}^T}{\underline{\ell}^T\underline{1}} \tag{6.6.3}$$

i.e., one obtains the unique probability vector that is the left eigenvector of $\mathbf{a}_G$.

With this conditioning, the probability that the process remains in G a subsequent time τ is, for $\underline{\ell}^T$ normalized,

$$\underline{\ell}^T e^{-\nu\tau[I-\mathbf{a}_G]}\underline{1}_G = e^{-\nu\tau(1-\theta_{\nu G})} = e^{-\alpha_{1G}\tau}. \tag{6.6.4}$$

The quasi-stationary exit time density from G is therefore the pure exponential

$$s_Q(\tau) = \alpha_{1G}e^{-\alpha_{1G}\tau}. \tag{6.6.5}$$

This may be stated as a limit theorem.

<u>Theorem 6.6.2</u>. Let N(t) be ergodic on $\mathscr{N} = G+B$ where G, a proper subset of $\mathscr{N}$, is irreducible. Then

$$\lim_{t\to\infty} P[N(t+\tau) \in G|N(t') \in G,\ 0 < t' < t] = e^{-\alpha_{1G}\tau},$$

where α_{1G} is the principal decay rate for the lossy process on G.

This theorem has validity in more elaborate process

settings than that of finite chains. It is a special case of a result, unpublished, with E. Arjas.† The theorem may be restated as a kind of metaprinciple. The longer an ergodic system has been working, the more exponential is the distribution of its residual lifetime, in the absence of other information.

We will complete the discussion of this section by proving the spectral representation (6.6.2). Let $\mathbf{a}_G = \theta_G J_G + \Delta_G$ as in §1.2 where $J_G = \underline{r}_G \underline{\ell}_G^T / \underline{\ell}_G^T \underline{r}_G$ is the idempotent dyad obtained from the positive left and right eigenvectors of $\mathbf{a}_G$, and Δ_G has a spectral radius strictly less than θ_G. Then

$$e^{-\nu t(I_G - \mathbf{a}_G)} = e^{-\nu t}\{I + \sum_1^\infty \frac{(\nu t)^k}{k!}(\theta_G J + \Delta)^k\}.$$

Hence as in §1.2

$$\begin{aligned} e^{-\nu t(I_G - \mathbf{a}_G)} &= e^{-\nu t}\{I + \sum_1^\infty \frac{(\nu t)^k}{k!}(\theta_G^k J + \Delta^k)\} \\ &= e^{-\nu t}\{(e^{\nu t\theta_G} - 1) J + e^{\nu t}\Delta\} \\ &= e^{-\nu t(1-\theta_G)}(J + \epsilon(t)), \quad \epsilon(t) \to 0, \ t \to \infty, \end{aligned}$$

giving the representation needed.

§6.7. The Connection Between Exit Times and Sojourn Times. A Renewal Theorem.

With the ergodic sojourn time defined as in §6.5, one denotes by $S_V(t) = \int_0^t s_V(\tau)d\tau$ its c.d.f. and by $\overline{S}_V(t) = 1 - S_V(t)$ its *survival function*. Between the ergodic sojourn

†More general results may be found in E. Arjas and E. Nummelin [1]. See also E. Nummelin [54], [55].

time and the ergodic exit time, a relationship holds, similar to the relation between the lifetime of a renewal process and the residual lifetime at stationarity. Specifically:

Theorem 6.7A. For any finite ergodic Markov chain $N(t)$ on $\mathscr{N} = G+B$, the ergodic exit time density $s_E(\tau)$ and $\bar{S}_V(\tau)$, the survival function for the ergodic sojourn time, have the simple relation

$$s_E(\tau) = \frac{\bar{S}_{V(\tau)}}{\mu_V},$$

where $\mu_V = \int_0^\infty \bar{S}_V(\tau)d\tau$ is the expected sojourn time. The ergodic sojourn time density and ergodic exit time density have the same relation as the lifetime and residual lifetime at ergodicity of an ordinary renewal process. In particular, the ergodic exit time distribution is absolutely continuous and monotone decreasing.

This theorem is important and useful, both conceptually and as a calculational tool. The following ingredients are needed for the proof:

I. Write $\sigma_n(s)$ for the Laplace transform of the passage time density $s_{nB}(t)$ from $n \in G$ to the set B. (Tradition forces us here to use the letter s in two different meanings.) Then, from the following continuity of probability argument similar to that for (6.2.2), one has for $n \in G$:

$$\sigma_n(s) = \frac{\nu_n}{\nu_n+s}\left(\sum_{k\in G} \frac{\nu_{nk}}{\nu_n}\sigma_k(s) + \sum_{m\in B} \frac{\nu_{nm}}{\nu_n}\right). \qquad (6.7.2)$$

The passage time from n to B starts with an exponential dwell time in state n (first factor), then the process moves either to some state in G, or to B (respective terms in the second factor).

II. Let $t \to \infty$ in the following differential equations

$$\frac{d}{dt} P\{N(t)=n\} = -\nu_n P\{N(t)=n\} + \sum_{m\in G\cup B} \nu_{mn} P\{N(t)=m\}. \quad (6.7.3)$$

Then, from stationarity, the left side vanishes, and $P\{N(t)=n\} \to e_n$, so that

$$e_n\nu_n = \sum_{m\in G+B} e_m\nu_{mn}. \quad (6.7.4)$$

Summing over all $n \in G$, and using $\nu_n = \sum_{m\in G+B} \nu_{nm}$ one obtains after cancellation

$$\sum_{n\in G}\sum_{m\in B} e_n\nu_{nm} = \sum_{n\in G}\sum_{m\in B} e_m\nu_{mn}. \quad (6.7.5)$$

Equation (6.7.4) says that for any ergodic chain there is balance between departures and arrivals for any state at stationarity. Equation (6.7.5) says that for any parition of the state space $\mathscr{N} = G + B$ there is balance at stationarity between departures from the set G and arrivals to G from the set B.

Proof of Theorem 6.7A. If we refer back to the definitions of $s_E(\tau)$ and $s_V(\tau)$ in (6.5.4) and (6.5.6) we see that one must prove

$$\frac{\sum_{n\in G} e_n s_{nB}(\tau)}{\sum_{n\in G} e_n} = \frac{\sum_{n\in G}\sum_{m\in B} e_m\nu_{mn}\int_t^\infty s_{nB}(\tau)d\tau}{\sum_{n\in G}\sum_{m\in B} e_m\nu_{mn}\mu_n}, \quad (6.7.6)$$

where $\mu_n = \int_0^\infty \tau\, s_{nB}(\tau)d\tau$. As will be shown, there will be equality in (6.7.6) between the numerators, and between the denominators of both sides. Equation (6.7.2) may be rewritten

$$s\,\sigma_n(s) = -\nu_n\sigma_n(s) + \sum_{k\in G} \nu_{nk}\sigma_k(s) + \sum_{m\in B} \nu_{nm}. \quad (6.7.7)$$

When we multiply by e_n and sum over $n \in G$, we find

$$s \sum_{n\in G} e_n\sigma_n(s) = - \sum_{n\in G} e_n\nu_n \sigma_n(s) + \sum_{n\in G}\sum_{k\in G} e_n\nu_{nk} \sigma_k(s)$$

$$+ \sum_{n\in G}\sum_{m\in B} e_n\nu_{nm}. \tag{6.7.8}$$

Use of (6.7.4) and (6.7.5) gives

$$s \sum_{n\in G} e_n\sigma_n(s) = \sum_{n\in G}\sum_{m\in B} e_m\nu_{mn} (1-\sigma_n(s)). \tag{6.7.9}$$

If we divide through by s and let $s \to 0+$, we see that

$$\sum_{n\in G} e_n = \sum_{n\in G}\sum_{m\in B} e_m\nu_{mn} \mu_n, \tag{6.7.10}$$

so that equality of the denominators of (6.7.6) is shown. Further, since $\frac{1}{s} (1-\sigma_n(s))$ is the Laplace transform of $\int_t^\infty s_{nB}(\tau)d\tau$, Eq. (6.7.9) also proves equality of the numerators. This comples the proof.

Because (6.7.10) can be written as

$$\sum_{n\in G} e_n = \sum_{n\in G}\sum_{m\in B} e_m\nu_{mn} \frac{\sum_{n\in G}\sum_{m\in B} e_m \nu_{mn}\mu_n}{\sum_{n\in G}\sum_{m\in B} e_m \nu_{mn}},$$

one finds in passing that

$$P(G) = i\, \mu_V, \tag{6.7.11}$$

where P is the ergodic probability, i the ergodic flow rate (§6.4) or asymptotic frequency of transition from G to B (or B to G) and μ_V the "expected sojourn time". This equation is of interest in its own right and is a handy tool. It may be obtained more heuristically from the following simple argument. Consider the indicator function $\phi_G(t)$ for the process, i.e., $\phi_G(t) = 1$, when $N(t) \in G$, $\phi_G(t) = 0$, when

$N(t) \in B$. For a sample path $n(\omega,t)$ of the process from $t = 0 \to \infty$ the time spent in G on $(0,t) \sim \mu_V N(t)$ where

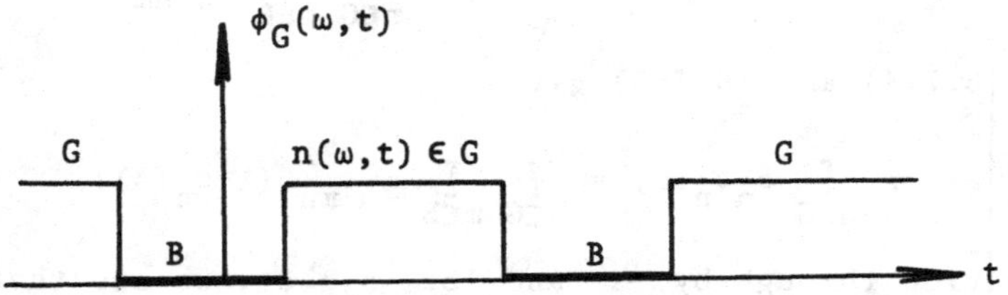

$N(t)$ is the number of entries into G on $(0,t)$. Then $P(G)$, the limiting fraction of time spent in G is

$$P(G) = \lim_{t\to\infty} \frac{N(t)\,\mu_V}{t} = i\,\mu_V. \tag{6.7.12}$$

A more elaborate derivation of (6.7.11) based on the ideas of semi-Markov processes and renewal theory provides a bridge to the justification of Theorem 6.7A and (6.7.11) for processes with richer state space structure. Given the decomposition of the state space $\mathscr{N} = G + B$, with alternating sojourns on G and B define a process $J(t)$, by $J(t) = n$ if n is the state where the current sojourn of $N(t)$ initiated. A simple examination shows that $J(t)$ is a semi-Markov process, and it is easily seen that the renewal density $h_{Bn}(t)$ for state $n \in G$ for the J-process equals $\Sigma_{m\in B}\, h_{mn}(t)$, where $h_{mn}(t)$ is the renewal density for state (m,n) in the T-process, defined in §6.4.

Hence from (6.4.4) and (6.4.7)

$$h_{Bn} = \lim_{t\to\infty} h_{Bn}(t) = \lim_{t\to\infty} \sum_{m\in B} h_{mn}(t) = \sum_{m\in B} e_m \nu_{mn}, \qquad (6.7.13)$$

and

$$i = h_{BG} = \sum_{n\in G} h_{Bn} = \sum_{n\in G} \sum_{m\in B} e_m \nu_{mn}. \qquad (6.7.14)$$

Denoting by $\bar{S}_{nB}(t)$ the survival function of $s_{nB}(t)$, i.e., $\bar{S}_{nB}(t) = P\{N(t') \in G \quad 0 < t' < t \; |N(0) = n\}$, $n \in G$, one has

$$P\{N(t)\in G, J(t) = n \mid N(0)\in B\}$$
$$= \int_0^t h_{Bn}(t-t')\bar{S}_{nB}(t')dt', \qquad (6.7.15)$$

where $h_{Bn}(t) = \Sigma_{m\in B}\, p_n(t)\nu_{mn}$.

Summing over $n \in G$ and letting $t \to \infty$, the left side of (6.7.15) tends to $P(G)$ and in the right side by the dominated convergence theorem, one obtains $\Sigma_{n\in G}\, h_{Bn}\, \mu_n$, where μ_n is defined as in (6.7.6).

Hence, by (6.7.13) $P(G) = \Sigma_{n\in G}\, \Sigma_{m\in B}\, e_m \nu_{mn}\, \mu_n$, which again is (6.7.10).

§6.8. A Comparison of the Mean Ergodic Exit Time and Mean Ergodic Sojourn Time for Arbitrary Chains.

From the basic relation (6.7.1) in Theorem 6.7A

$$s_E(\tau) = \frac{\bar{S}_V(\tau)}{E\,T_V}, \qquad (6.8.1)$$

one finds for the moments of the ergodic exit time T_E and the ergodic sojourn time T_V

$$ET_E = \int_0^\infty \tau\, s_E(\tau)d\tau = \frac{1}{ET_V}\int \tau\, \bar{S}_V(\tau)d\tau = \frac{ET_V{}^2}{2ET_V}. \qquad (6.8.2)$$

If the exit time has a first moment, the sojourn time must

have a second moment. From (6.8.2) it follows that

$$\frac{ET_E}{ET_V} = \frac{ET_V^2}{2(ET_V)^2} = \frac{1}{2}\left(1 + \left(\frac{\sigma^2}{\mu^2}\right)_{T_V}\right). \tag{6.8.3}$$

From $\frac{\sigma^2}{\mu^2} \geq 0$, one finds the inequality

$$ET_E > \frac{1}{2} ET_V = \frac{1}{2} \frac{P(G)}{i}. \tag{6.8.4}$$

There seems to be no intuitive argument supporting this inequality. The equation is of practical importance because it provides a lower bound for ET_E in terms of ET_V which is available directly from the ergodic probabilities and transition rates.

A bound on $\left(\frac{\sigma^2}{\mu^2}\right)_{T_E}$ for general chains is available from the following argument. Theorem 6.7A assures that $s_E(\tau)$ is monotone decreasing. It follows that T_E has a representation of the form [39]

$$T_E = WU, \tag{6.8.5}$$

where W is a positive mixing random variable, U has uniform density $f_U(x) = 1$ on $(0,1)$, and W and U are independent. The relation (6.8.5) is valid in the sense that the distribution of T_E is a mixture of scaled uniforms. W and U are not random variables in the conventional setting of a probability space. From (6.8.5) we have

$$\frac{ET_E^2}{(ET_E)^2} = \frac{EW^2}{(EW)^2}\,\frac{4}{3} \geq \frac{4}{3}, \tag{6.8.6}$$

so that

$$\left(\frac{\sigma^2}{\mu^2}\right)_E \geq 1/3. \tag{6.8.7}$$

It can be shown that for finite Markov chains and partitions thereof the value 1/3 is an infimum.

We will see in the next section that for time-reversible chains, one has

$$E\, T_E \geq E\, T_V, \tag{6.8.8}$$

and that

$$\left(\frac{\sigma^2}{\mu^2}\right)_E \geq 1; \quad \left(\frac{\sigma^2}{\mu^2}\right)_V \geq 1. \tag{6.8.9}$$

§6.9. Stochastic Ordering of Exit Times of Interest for Time-Reversible Chains.

For ergodic time-reversible processes, such as the birth-death processes and tree processes of §2.5 and the systems of independent Markov components discussed there and in Example 6.1A, the sojourn times and exit times have a simple structural form, that of complete monotonicity. Specifically we have:

Theorem 6.9A. Let $N(t)$ be an ergodic finite time-reversible Markov chain on $\mathcal{N}$. Let $\mathcal{N} = G+B$, $GB = \emptyset$ be an arbitrary partition of $\mathcal{N}$. Then the ergodic exit time from G and ergodic sojourn time on G have completely monotone densities.

Proof: We may assume without loss of generality that $\mathfrak{a}_G$ is irreducible. For the property of complete monotonicity is preserved under limits, and one may consider $\mathfrak{a}_G$ as the limit of a sequence of such having the irreducibility needed. When $\mathfrak{a}_G$ is irreducible, the matrix $\mathfrak{e}_{DG}\, \mathfrak{p}_G(t) = \mathfrak{e}_{DG} \exp\{-\nu t[\mathfrak{I}_G - \mathfrak{a}_G]\}$ has the symmetric spectral representation (3.3.8)

$$\mathfrak{e}_{DG}\, \mathfrak{p}_G(t) = \sum_1^{N_G} e^{-\alpha_j \tau}\, \underline{w}_G^{(j)}\, \underline{w}_G^{(j)T}, \tag{6.9.1}$$

where $\alpha_j > 0$, and $\underline{w}_G^{(j)}$ are real. Then

$$\sum_G \sum_G e_m \, p_{G;mn}(t) = \underline{1}_G^T \, e_{DG} \, p_G(t) \underline{1}_G = \sum_1^{N_G} e^{-\alpha_j \tau} (\underline{1}_G^T \, \underline{w}^{(j)})^2 . \tag{6.9.2}$$

But (6.9.2), apart from normalization, is the survival function $\overline{S}_E(\tau)$ of the ergodic exit time, and from (6.7.2) we see that this is completely monotone. The density $S_E(\tau) = -\frac{d}{dt}\overline{S}_E(\tau)$ is then also completely monotone and the ergodic sojourn time density is completely monotone as well from Theorem 6.7A. This completes the proof.

The demonstration of (6.8.8) for time-reversible chains now follows from Theorem 6.9A. For as seen in Theorem 5.6A, the complete monotonicity of $s_V(\tau)$ implies that $(\sigma^2/\mu^2)_V \geq 1$ and (6.8.8) is obtained from (6.8.3). Not only do we have $ET_E \geq ET_V$ (with equality only for pure exponentiality), but, more strongly,

$$ET_E^{\,n} \geq ET_V^{\,n}. \tag{6.9.3}$$

Indeed an even stronger result is available. A definition is needed.

<u>Definition 6.9B</u>. Let X, Y be random variables. The r.v. X is said to be stochastically larger than the r.v. Y when $\overline{F}_X(x) \geq \overline{F}_Y(x)$, i.e., when

$$P[X > x] \geq P[Y > x] \quad \text{all real } x. \tag{6.9.4}$$

In this case we write $X \succ Y$.

<u>Theorem 6.9C</u>. Let $N(t)$ be a finite ergodic time-reversible chain and let $\mathcal{N} = G+B$, $GB = \emptyset$ be any partition of its state space. Let T_Q, T_E, and T_V be the quasi-stationary exit time, ergodic exit time and ergodic sojourn time for set G. Then

$$T_Q \succ T_E \succ T_V. \tag{6.9.5}$$

Proof: The reasoning required to prove (6.9.5) is based on simple notions of renewal theory. Suppose the lifetime of a renewal process is DHR† and has density $a(x)$. Then as discussed in §5.9, $N(x+y) - N(x)$ decreases with x, and $\bar{A}(x+y)/\bar{A}(x)$ increases with x for every y. This then implies that T_x the residual life given survival to age x increases stochastically with x in the sense of Def. 6.9B i.e., $T_x \succ T_y$ when $x \geq y$. The residual lifetime R at ergodicity is a mixture of the lifetimes T over different ages x. It follows that $R \succ T_0$. But from Theorem 6.9C, T_E is related in distribution to T_V as R is to T_0, this being the content of Theorem 6.7A. For the time-reversible case, moreover, T_V is DHR. We then have $T_E \succ T_V$. In the same context, however, T_Q is the limit in distribution of T_x, and hence is larger stochastically than every T_x. It follows that $T_Q \succ T_E \succ T_V$ as stated.

A fourth exit time is often of interest in special cases. Two examples illustrate what is involved.

Example 6.9D. Consider an ergodic truncated birth-death process on $\mathcal{N} = \{0,1,2,\ldots.K\}$ and let $\mathcal{N} = G+B$ where $G = \{0,1,2,\ldots.R\}$ with $R < K$. Another exit time is the time from state 0 to set B. Clearly $T_{0B} \succ T_{1B} \succ T_{2B} \ldots \succ T_{RB}$. Hence T_{0B} is stochastically larger than every exit time from G, since these are just mixtures of T_{mB}. In particular the sojourn time T_V on G is T_{RB}. Since birth-death processes are time-reversible (§2.5), we

† The designation DFR for decreasing failure rate is common in the literature.

have for this case

$$T_{0B} \succ T_Q \succ T_E \succ T_V. \qquad (6.9.6)$$

An example of this in reliability is the system of K identical Markov components required to work for the system to work.

<u>Example 6.9E</u>. Consider any system of k independent Markov components of the kind discussed in Example 6.3A. Here the state of key interest is the perfect state $\underline{1}$ where all components are working. It has been shown by S. Ross that if the system is coherent, its failure time distribution from the perfect state is NBU, i.e., new better than used. Indeed his method of proof shows that $T_{\underline{1}B}$ is stochastically larger than $T_{\underline{m}B}$ for every working, i.e., for all $\underline{m}$ in the good set G. It follows that

$$T_{\underline{1}B} \succ T_Q \succ T_E \succ T_V$$

for such systems.

§6.10. Superiority of the Exit Time as System Failure Time; Jitter.

In a reliability context, such as the situation described in Example 6.5A, the exit time T_E is often a more appropriate failure time for the system than the sojourn time T_V, for the following reason. T_V is very sensitive to properties of boundary states, and as a consequence its distribution may have a mass point at 0. An example of such is provided by an ergodic Ornstein-Uhlenbeck diffusion process on the real halfline with $G = [0,L)$, $B = [L,\infty)$. Since sample paths of this process are "infinitely fuzzy", a fraction of the sojourns on G have zero duration. In contrast, the ergodic exit time from G for such process is well defined.

The same undesirable behavior of the ergodic sojourn time is illustrated in the following example. Consider a birth-death process $N(t)$ on $\mathcal{N} = \{0,1,2,\ldots\}$ and a sequence of partitions $\mathcal{N} = G_N \cup B_N$, where $G_n = \{0,1,\ldots,N\}$. $N = 1,2,\ldots$ Suppose the process is ergodic. For each N, denote by T_{EN} and T_{VN} the ergodic exit time from G_N, and the ergodic sojourn time in G_N respectively. Then it can be shown that T_{EN} is monotone in N in the following sense

$$P(G_N)\cdot ET_{EN} \quad \text{increases with } N, \quad P(G_N) \uparrow 1. \qquad (6.10.1)$$

This follows from

$$P(G_N)\ ET_{EN} = \sum_{n\in G_N} e_n \int_0^\infty p_n^{(*N)}(\tau)d\tau, \qquad (6.10.2)$$

where $p_n^{(*N)}(\tau)$ is the survival probability for the transient chain on G_N, given the process starts in $n \in G$. Precisely: $p_n^{(*N)}(t) = P\{N(\tau) \in G_N \ \ 0 \le \tau \le t \mid N(0) = n\}$. In (6.10.2) not only the number of terms increases with N, but in each term $p_n^{(*N)}(t)$ increases in N. In contrast with the monotonicity property (6.10.1) for the ergodic exit time, the ergodic sojourn time does not have such behavior. In fact, it is possible to construct a birth-death process in which any finite sequence of expected sojourn times $\{E\ T_{VN}\}_{N=1}^{L}$ can be an arbitrary specified sequence $\{\alpha_N\}_{N=1}^{L}$ of positive numbers. This property, demonstrated below, could make the use of the ergodic sojourn time as a system failure time less desirable.

Note that, in any time-reversible chain (hence in any birth-death process) the ergodic probabilities are determined by the potential coefficients (§3.3) and hence by the ratio of upward and downward hazard rates between neighboring states. Hence, multiplying both upward and downward hazard rates between states N and N+1 (i.e., λ_N and μ_{N+1} respectively), by a factor K does not change the ergodic probabilities.

Now, suppose that a process $N(t)$ is given with $E\,T_{V\ell} = \alpha_\ell$ for $\ell = 1,2,\ldots N$. It will be shown that a process $N^*(t)$ can be found with $E\,T_{V\ell} = \alpha_\ell$ for $\ell = 1,2,\ldots,N, N+1$. Specializing (6.7.11) to a birth-death process, one finds

$$E\,T_{V\ell} = \frac{P(G_\ell)}{i_{B_\ell G_\ell}} = \frac{\sum_{n=1}^{\ell} e_n}{e_{\ell+1}\mu_{\ell+1}}\,. \tag{6.10.3}$$

$N^*(t)$ (governed by $\{\lambda_n^*, \mu_n^*\}$) is now obtained from $N(t)$, by taking

$$\lambda_{n-1}^* = \lambda_{n-1},\ \mu_n^* = \mu_n \quad n = 1,2,\ldots,N, N+2, N+3,\ldots,L.$$
$$\lambda_N^* = K\,\lambda_N,\ \mu_{N+1}^* = K\mu_{N+1}.$$

Then it is easily seen from (6.10.3) that $E\,T_{V\ell}^* = ET_{V\ell}$, $\ell = 1,2,\ldots,N-1, N+1,\ldots,L$, but $ET_{VN}^* = \frac{1}{K}\,ET_{VN}$. An appropriate choice of K will put $E\,T_{VN}^* = \alpha_N$. The ergodicity of all processes is assured since no hazard rates beyond state L are altered, and single step downward passage time beyond L are unchanged.

Chapter 7

The Fundamental Matrix, and Allied Topics

§7.00. Introduction.

In systems with a reasonably small number of components, where machine computation is feasible, techniques are needed to calculate explicitly the mean and variance of the passage time for the Markov chain $N(t)$ from a state to a set B. When the entry set B^* is much smaller than B, the rank of the problem reduces to that of B^* provided that the transition probabilities $p_{mn}(t)$ are known. In such calculations there is a fundamental matrix Z, the anologue of that in discrete time, whose properties are of interest. The fundamental matrix is related to the covariance function for stationary processes defined on the chain $N(t)$. It also plays a key role in the central limit theorem for additive processes defined on the chain, in ergodic potential theory,† and in perturbation theory for Markov chains.

Concluding sections of the chapter establish structure

†The role of the fundamental matrix in the potential theory of ergodic Markov chains has been developed recently. See [2], [41], [62], and [63]. For perturbation theory see Syski [62].

needed for Chapter 8, where rarity and exponentiality are examined. A triangle inequality for mean passage times is of particular importance.

§7.1. The Fundamental Matrix for Ergodic Chains.

In the theory of ergodic finite Markov chains in discrete time, one encounters the so-called fundamental matrix [46] defined by

$$Z_a = [I-a+\underline{1}\,\underline{e}^T]^{-1} = \sum_{k=0}^{\infty} (a-\underline{1}\,\underline{e}^T)^k, \tag{7.1.1}$$

where $\underline{e}^T$ is the ergodic vector for the chain governed by a. The convergence of the series in (7.1.1) is assured since the spectral radius of $(a-\underline{1}\,\underline{e}^T)$ is less than one, a consequence of the Perron-Romanovski-Frobenius Theorem. Since $(a-\underline{1}\,\underline{e}^T)^k = a^k-\underline{1}\,\underline{e}^T$, Z_a may be rewritten in the form

$$Z_a = I + \sum_1^{\infty}(a^k-\underline{1}\underline{e}^T), \tag{7.1.2}$$

so that

$$Z_{a;mn} = \delta_{mn} + \sum_1^{\infty}(a_{mn}^{(k)} - e_n). \tag{7.1.3}$$

Clearly Z_a, a, and $\underline{1}\underline{e}^T$ commute. When a has a complete set of eigenvectors, i.e., is diagonalizable, it has the spectral representation $a = \sum_1^N \lambda_j J_j$ in terms of its eigenvalues λ_j and orthogonal idempotent dyads J_j; Z_a has the corresponding representation

$$Z_a = J_1 + \sum_2^N \frac{1}{1-\lambda_j} J_j; \quad J_1 = \underline{1}\underline{e}^T. \tag{7.1.4}$$

For ergodic chains, the fundamental matrix Z_a plays the same role as the Green potential of §4.1 $g_a = \sum_0^{\infty} a^k$ does for transient or lossy chains. The removal of $\underline{1}\underline{e}^T$ from the summands of (7.1.1) is required for convergence.

A simple analogue of the fundamental matrix is available for finite ergodic chains in continuous time.

<u>Definition 7.1A</u>. Let $N(t)$ be any finite ergodic Markov chain in continuous time with transition probability matrix $\mathfrak{p}(t)$. The matrix

$$Z = \int_0^\infty \{\mathfrak{p}(t) - \underline{1}\underline{e}^T\} dt, \qquad (7.1.5)$$

will be called the fundamental matrix for $N(t)$.

We note that convergence of the integral is assured, since for finite ergodic chains $\mathfrak{p}(t) - \underline{1}\underline{e}^T$ always has an exponential tail. Indeed if $\mathfrak{p}(t) = \exp\{-\nu t[I - \mathfrak{a}_\nu]\}$ one has from $\mathfrak{a}_\nu = \underline{1}\underline{e}^T + \Delta = J_1 + \Delta$, via the procedure of §6.6,

$$\mathfrak{p}(t) - \underline{1}\underline{e}^T = -e^{-\nu t}(\underline{1}\underline{e}^T) + e^{-\nu t} e^{\nu t \Delta}. \qquad (7.1.6)$$

The spectral radius of $e^{\nu t \Delta}$ is $e^{\nu t p}$ where p is the spectral radius of Δ with $0 \le p < 1$. Hence the spectral radius of $(\mathfrak{p}(t) - \underline{1}\underline{e}^T)$ is $e^{-\nu t(1-p)}$ as stated. It follows from (7.1.5) and (7.1.6) that

$$\nu Z = [I - \mathfrak{a}_\nu + \underline{1}\underline{e}^T]^{-1} - \underline{1}\underline{e}^T. \qquad (7.1.7)$$

This enables one to calculate Z_{mn} via computer inversion programs when the number of states is not too large. The first term on the right is the fundamental matrix (7.1.1) for the discrete time chain governed by $\mathfrak{a}_\nu$.

The matrix Z arises constantly in the analysis of chains. As we will see, it provides a tool for the evaluation of the mean passage time from a state to a set and of associated ruin probabilities. It also appears, as we will see, in the asymptotic variance of the central limit theorem for additive processes defined on a chain, and arises in the covariance function for a process $f(N(t))$, where $f(x)$ is given and $N(t)$ is a chain of interest.

Closely related to the fundamental matrix is the matrix Laplace transform

$$\zeta(s) = \int_0^\infty e^{-st}(\mathbb{P}(t)-J_1)dt, \qquad (7.1.8)$$

where $\mathbb{P}(t)$ is the transition probability matrix for a finite ergodic chain $N(t)$ and J_1 is the principal dyad $\underline{1}\underline{e}^T$. Then from the spectral representation (3.2.4) the components of $\zeta(s)$ have a common set of simple poles on the negative real axis and are regular everywhere else, including the origin $s = 0$. Hence the power series

$$\zeta(s) = \sum_0^\infty \frac{s^k}{k!} (-1)^k Z_k, \qquad (7.1.9)$$

has components which are absolutely convergent in some circle of convergence with center at $s = 0$, and

$$Z_k = \int_0^\infty t^k(\mathbb{P}(t)-J_1)dt. \qquad (7.1.10)$$

We note that Z_0 is the fundamental matrix.

The matrices Z_k are simply related to Z_0. Assume for the moment that $\mathbb{P}(t)$ has a complete set of eigenvectors. Then $\mathbb{P}(t) = J_1 + \sum_2^N e^{-\alpha_k t} J_k$, and from (7.1.10),

$$Z_k = \sum_2^N \frac{k!}{(\alpha_j)^{k+1}} J_j. \qquad (7.1.11)$$

It follows immediately that

$$Z_k = k!\, Z_0^{k+1} \quad k = 1,2,\ldots . \qquad (7.1.12)$$

From (7.1.9) one then has

$$\zeta(s) = Z_0(I+s\,Z_0)^{-1}. \tag{7.1.13}$$

The non-availability of a complete set of vectors may be avoided by a slightly more elaborate argument based on the identity

$$\zeta(s) = [(s+\nu)I-\nu a_\nu+\nu\,J_1]^{-1} - (s+\nu)^{-1}\,J_1, \tag{7.1.14}$$

obtained from (7.1.6) via integration. If we post-multiply by the term in square brackets, differentiate with respect to s and set $s = 0$, (7.1.12) is obtained by induction.

§7.2. The Structure of the Fundamental Matrix for Time-Reversible Chains.

Consider a finite ergodic chain in discrete time N_k which is reversible in time, and governed by the stochastic matrix a. For such chains the slightly hidden self-adjointness of a and Z_a assures a full set of real eigenvectors, and the representation (7.1.4) is assured. Clearly the eigenvalues of Z_a are then positive, and the J_j real. Moreover, just as $e_D^{1/2}\,a e_D^{-1/2}$ is symmetric for time-reversible chains (cf. §1.3) so also is $e_D^{1/2}\,Z_a\,e_D^{-1/2}$. It follows that:

Prop. 7.2A. If N_k is a finite ergodic time reversible chain, and Z_a is its fundamental matrix (7.1.1), then $e_D^{1/2}\,Z_a\,e_D^{-1/2}$ is positive definite. For any finite time-reversible ergodic chain in continuous time, the spectral representation (3.2.4) with $\alpha_1 = 0$, $J_1 = \underline{1}\underline{e}^T$ and (7.1.5) gives†

†The reader is urged to compare (7.1.4) for discrete time chains with (7.2.1) for continuous time chains.

$$Z = \sum_{j=2}^{N} \frac{1}{\alpha_j} J_j, \tag{7.2.1}$$

with α_j positive and J_j real. One then has the following continuous time analogue of Prop. 7.2A.

Prop. 7.2B. For any finite ergodic time-reversible chain $N(t)$ in continuous time with transition matrix $p(t)$, the fundamental matrix Z_0 defined by (7.1.5) is singular and $e_D^{1/2} Z_0 e_D^{-1/2}$ is positive semi-definite.

In the study of the passage time to a set, the following question arises: Suppose G is a proper subset of the state space $\mathcal{N}$ for a given finite ergodic time-reversible chain. Let Z_G be the principal submatrix obtained from Z by deleting rows and columns for state indices not in G. Is the matrix Z_G non-singular? The answer is yes.

Theorem 7.2C. Let $N(t)$ be finite, ergodic, and time-reversible on $\mathcal{N}$. Let G be a proper subset of $\mathcal{N}$ and let Z_G be the principal submatrix associated with set G. Then Z_G is non-singular.

Proof: Let $\underline{x} \in R^N$, and let $g(\underline{x}) = \underline{x}^T e_D^{1/2} Z e_D^{-1/2} \underline{x}$. From (7.1.6) Z is of rank $N-1$, $g(\underline{x}) \geq 0$ for $\underline{x}$ real, and the only $\underline{x}$ for which $g(\underline{x}) = 0$ is $\underline{x} = 0$, and $\underline{x} = K \underline{e^{1/2}}$, i.e., $x_n = K e_n^{1/2}$ with K constant. Let $\underline{y} \in R^N$ with $y_n = 0$, $n \notin G$. Then $g(\underline{y}) = 0$ only if $\underline{y} = 0$. But the set $\{g(\underline{y})\}$ so obtained is the set of quadratic form values of $(e_D^{1/2})_G Z_G (e_D^{-1/2})_G$ where $(e_D^{1/2})_G$ is the diagonal matrix restricted to G. Hence $(e_D^{1/2})_G Z_G (e_D^{-1/2})_G$ is positive definite and Z_G is non-singular.

Similar considerations apply to the higher order

matrices Z_k of (7.1.10). We observe from (3.2.4) and (7.1.10) that

$$e_D^{1/2} Z_k e_D^{-1/2} = \sum_{j=2}^{N} (e_D^{1/2} J_j e_D^{-1/2}) \int_0^\infty t^k e^{-\alpha_j t}\, dt. \qquad (7.2.2)$$

It follows as in §3.2 that $e_D^{1/2} Z_k e_D^{-1/2}$ is positive semi-definite, its null space is of rank 1, and the eigenvectors of Z_k corresponding to this null space are $\underline{e}^T$ and $\underline{1}$. The situation is precisely as in Theorem 7.2C. If we deal with a principal submatrix of Z_k, say $(Z_k)_G$ for the set G, then $(e_D^{1/2} Z_k e_D^{-1/2})_G$ is positive-definite, the reasoning being identical to that of Theorem 7.2C. Formally:

<u>Theorem 7.2D</u>. Let Z_k be defined by (7.1.11) in the setting of Theorem 7.2C. Then if G is any proper subset of N, the matrices Z_{kG} $k = 0,1,2,\ldots$ are all non-singular, and have positive eigenvalues. One last theorem is needed.

<u>Theorem 7.2E</u>. Let $N(t)$ be a finite ergodic chain, with transition probability matrix $p(t)$, and Laplace transform $\pi(s)$. Let $\pi_A(s)$ be a principal submatrix of $\pi(s)$ obtained by deleting rows and columns with indices not in A. Then $\pi_A(s)$ is non-singular for $s > 0$.

<u>Proof</u>: With the aid of uniformization (Chapter 2) we may write for $m, n \in A$

$$\frac{d}{dt} p_{mn}(t) = -\nu p_{mn}(t) + \nu \sum_{r \in A} p_{mr}(t) * s_{rn}(t), \qquad (7.2.3)$$

where the asterisk denotes convolution and $s_{rn}(t)$ is the n'th component for the vector passage time density back to the set after a departure from state r. Since transitions within the set may occur, the density is a generalized density with some components having support near $t = 0$. Then

$$s\pi_{mn}(s) - \delta_{mn} = -\nu\, \pi_{mn} + \nu \sum_{n\in A} \pi_{mr}\, \sigma_{mn}(s).$$

In matrix form

$$[(s+\nu)I - \nu\sigma(s)]\pi(s) = I. \qquad (7.2.4)$$

For $s > 0$, the matrix $\frac{\nu}{\nu+s}\,\sigma(s)$ is strictly substochastic, and the matrix in square brackets on the left is non-singular. It follows that $\pi(s)$ is non-singular. This is a simple version of a proof given earlier [42].

The gist of the argument is simplified when G has only one state. The equation $\frac{dp}{dt} = -\nu p(t) + \nu p(t) * s(t)$ where s is the return time density is well known, and has a simple interpretation.

§7.3. Mean Failure Times and Ruin Probabilities for Systems with Independent Markov Components and More General Chains.[†]

Consider a system of K independent Markov components of the type discussed in Example 6.3A and Example 6.5A. The vector process $\underline{N}(t)$, whose j'th component $N_j(t)$ is 0 or 1 depending on whether the system component j is down or up, is a time-reversible ergodic Markov chain with 2^K states in its state space $\mathscr{N}$. Let $\mathscr{N} = G+B$ where B is the set of failure states. Suppose that B contains a small set of boundary states B^* of interest and one wishes to find the mean time $ET_{\underline{1}B^*}$ from the perfect state $\underline{1}$ to the set B^*, and the probability R_m, $m \in B^*$, for reaching set m before

[†]This section may be omitted without disturbing the continuity. It is of interest to those who must calculate passage times from a state to a set explicitly.

the other states of B^*. We will demonstrate a method that exploits the two simplifying elements in the problem a) the independence of the components; b) the relatively small number of states of interest.

To find the quantities of interest we consider the modified process $N^*(t)$ in which samples of $N(t)$ reaching B^* are replaced immediately at the perfect state $\underline{1}$. This corresponds to immediate repair of systems that fail. Let $p^*_{\underline{1}\underline{n}}(t) = P[\underline{N}^*(t) = \underline{n}|\underline{N}^*(0) = \underline{1}]$ and let $p_{\underline{1}\underline{n}}(t)$ be the corresponding transition probability for the unmodified process, i.e., $p_{\underline{1}\underline{n}}(t) = P[\underline{N}(t) = n|\underline{N}(0) = \underline{1}]$. Then as in §4.3, and §4.5 the distribution for the modified process will be obtained from that for the unmodified process by continually injecting negative mass at the states of B^* and positive mass at the state $\underline{1}$ in such quantity as to keep $p^*_{\underline{1}\underline{m}}(t) = 0$ for $\underline{m} \in B^*$ and simulate thereby the actual transferral. One then has for $\underline{N}^*(0) = \underline{N}(0) = \underline{1}$

$$p^*_{\underline{1}\underline{n}}(t) = p_{\underline{1}\underline{n}}(t) - \sum_{B^*} i_{\underline{m}}(t) * p_{\underline{m}\underline{n}}(t) + \left(\sum_{B^*} i_{\underline{m}}(t)\right) * p_{\underline{1}\underline{n}}(t); \quad \text{all } \underline{n} \in \mathscr{N}, \tag{7.3.1}$$

and

$$p^*_{\underline{1}\underline{n}}(t) = 0; \quad \underline{n} \in B^*. \tag{7.3.2}$$

From (7.3.1) and (7.3.2) one has for the Laplace transforms, denoting the perfect state by 0 and dropping the bars)[†]

$$\pi_{0n}(s) = \sum_{B^*} \tilde{i}_m(s)\, \pi_{mn}(s) - \{\sum_{B^*} \tilde{i}_m(s)\}\, \pi_{0n}(s), \tag{7.3.3}$$

[†]The remainder of this section applies to passage times and ruin probabilities from a state 0 to a set B^* for arbitrary chains.

for all $n \in B^*$. This is a square array of equations with $\tilde{i}_m(s)$ the unknowns, and $\pi_{mn}(s)$ and $\pi_{0n}(s)$ knowns. It is helpful to transcribe for comparison the equation set (7.2.4) for the vector passage time densities $(s_{0n}(t))$ from 0 to B^* and the transform set

$$p_{0n}(t) = \sum_{B^*} s_{0m}(t) * p_{mn}(t); \quad n \in B^* \tag{7.3.4}$$

$$\pi_{0n}(s) = \sum_{B^*} \sigma_{0m}(s)\, \pi_{mn}(s); \quad n \in B^*. \tag{7.3.5}$$

Equation (7.3.5) can be rewritten in vector notation as

$$\underline{\pi}_0^T(s) = \underline{\sigma}_0^T(s)\, \pi(s). \tag{7.3.6}$$

(In the remainder of this section, subscripts not explicitly written will be assumed to run over the set B^*. Thus (7.3.6) could be written as $\underline{\pi}_{0B^*}^T(s) = \underline{\sigma}_{0B^*}^T(s)\pi_{B^*B^*}(s)$. This convention should impose no hardship on the reader, and greatly simplifies the notation.)

From Theorem 7.1E and (7.3.6), we have

$$\underline{\sigma}_0^T(s) = \underline{\pi}_0^T(s)\, \pi^{-1}(s). \tag{7.3.7}$$

Equation (7.3.3) gives

$$\underline{\pi}_0^T(s) = \underline{\tilde{i}}^T(s)\, \pi(s) - [\underline{\tilde{i}}^T(s)\underline{1}]\ \underline{\pi}_0^T(s). \tag{7.3.8}$$

This can be written as

$$\underline{\tilde{i}}^T(s) = \underline{\pi}_0^T(s)\, \pi^{-1}(s)(1+\tilde{i}(s)), \tag{7.3.9}$$

where $\tilde{i}(s) = \underline{\tilde{i}}^T(s)\underline{1}$. Multiplying (7.3.9) by $\underline{1}$, we have

$$\frac{\tilde{i}(s)}{1+\tilde{i}(s)} = \underline{\pi}_0^T(s)\, \pi^{-1}(s)\underline{1}, \tag{7.3.10}$$

so that from (7.3.7),

$$\frac{\tilde{i}(s)}{1+\tilde{i}(s)} = \sigma(s), \tag{7.3.11}$$

where $\sigma(s) = \underline{\sigma}_0^T(s)\underline{1}$. Equation (7.3.11) confirms the renewal relationship expected between the quantities $\tilde{i}(s)$ and $\sigma(s)$. From (7.3.9) and (7.3.7)

$$\underline{\tilde{i}}^T(s) = \underline{\sigma}_0^T(s)\ [1+\tilde{i}(s)], \tag{7.3.12}$$

so that, with $s\ \underline{\tilde{i}}^T(s) \to \underline{i}_\infty^T$ as $s \to 0+$,

$$\underline{\sigma}_0(0) = \frac{\underline{i}_\infty}{i_\infty}. \tag{7.3.13}$$

We note that $[\underline{\sigma}_0(0)]_m = R_m$ the probability of reaching state $m \in B^*$ before reaching another state in B^*, given that the initial state is 0. We then have

$$R_m = \frac{i_m}{\Sigma i_k}. \tag{7.3.14}$$

There remains the calculation of i_m. For this purpose we rewrite (7.3.8) in the form

$$s\ \underline{\pi}_0^T(s) = s\ \underline{\tilde{i}}^T(s)[\pi(s) - \frac{\underline{1}\underline{e}^T}{s}] + s\underline{\tilde{i}}^T(s)\underline{1}[\frac{\underline{e}^T}{s} - \underline{\pi}_0^T(s)] \tag{7.3.15}$$

and pass to the limit $s = 0$. This gives

$$\underline{e}^T = \underline{i}_\infty^T\theta - \underline{i}_\infty^T\ \underline{1}\ \underline{\Theta}_0^T. \tag{7.3.16}$$

Here θ is the restricted fundamental matrix of §7.2, i.e., $\theta = \int_0^\infty (\mathrm{p}(t) - \underline{1}\underline{e}^T)dt = Z$ restricted to B^*. From (7.3.16) and Theorem (7.2C) we obtain the asymptotic renewal density for the replacements at $\underline{0}$, i.e.,

$$i_\infty = \frac{\underline{e}^T\theta^{-1}\underline{1}}{1-\underline{\Theta}_0^T\theta^{-1}\underline{1}}. \tag{7.3.17}$$

Here $\theta^{-1} = [Z_{B^*B^*}]^{-1}$. The denominator cannot be zero for $\underline{e}^T\theta^{-1}\underline{1}$ is positive by virtue of $\underline{1}^T e_D \theta^{-1}\underline{1} > 0$. We now have the following key results.

$$\mu_{OB^*} = E[T_{OB^*}] = \frac{1}{i_\infty} = \frac{1-\underline{\Theta}_0^T\theta^{-1}\underline{1}}{\underline{e}^T\theta^{-1}\underline{1}} \tag{7.3.18}$$

$$\underline{i}_\infty = \underline{e}^T\theta^{-1} + \frac{\underline{e}^T\theta^{-1}\underline{1}}{1-\underline{\Theta}_0^T\theta^{-1}\underline{1}}\,\underline{\Theta}_0^T\theta^{-1} \tag{7.3.19}$$

$$R_m = \frac{(\underline{i}_\infty)_m}{i_\infty}. \tag{7.3.20}$$

For the special case for which the set B^* consists of a single state n, we have from (7.3.18) for the mean time from state 0 to state n

$$\mu_{0n} = \frac{Z_{nn} - Z_{0n}}{e_n}. \tag{7.3.21}$$

The reader may wish to verify that (7.3.18) and (7.3.20) may be obtained directly from (7.3.6) with the help of Z. To see this, and also to calculate higher moments of the passage time, we proceed as follows.

Equation (7.3.6) may be written in the form

$$\pi_{0n}(s) - \frac{e_n}{s} = \sum_{B^*}\sigma_{0r}(s)\{\pi_{rn}(s) - \frac{e_n}{s}\} - \frac{e_n}{s}\,(1-\sum_{B^*}\sigma_{0r}(s)) \quad \text{for } n \in B^*. \tag{7.3.22}$$

The previous equation can be written in vector notation restricted to the set B^* to obtain

$$\underline{\zeta}_0^T(s) = \underline{\sigma}_0^T(s)\,\zeta(s) - \frac{\{1-\sigma_0(s)\}}{s}\,\underline{e}^T. \tag{7.3.23}$$

where $\zeta(s)$ represents the transform of $p(t) - \underline{1}\underline{e}^T$ as discussed in §7.1.

Differentiation of equation (7.3.23) leads to moments of the passage time density. Specifically from evaluation at $s = 0$ we have

$$\underline{\Theta}_0^T = \underline{R}^T \theta_0 - \underline{e}^T T_1, \tag{7.3.24}$$

and differentiation of (7.2.23) at $s = 0$ yields

$$\underline{\Theta}_1^T = \underline{\mu}_1^T \theta_0 + \underline{R}^T \theta_1 + \underline{e}^T \frac{T_2}{2}. \tag{7.3.25}$$

Here $(\underline{\Theta}_0^T)_n = (Z_0)_{0n}$; $(\underline{\Theta}_1^T)_n = (Z_1)_{0n}$; $R_n = \int_0^\infty s_{0n}(\tau)d\tau$; $\mu_{1n} = \int_0^\infty \tau s_{0n}(\tau)d\tau$; T_1 and T_2 are the first and second moments of $s_0(\tau) = \Sigma\, s_{0n}(\tau)$, the passage time density of interest. We note that $\underline{R}^T\underline{1} = 1$ and $\underline{\mu}_1^T\underline{1} = T_1$.

Post-multiplying (7.3.24) by θ_0^{-1}, one again obtains (7.3.18)

$$T_1 = \frac{1 - \underline{\Theta}_0^T \theta_0^{-1} \underline{1}}{\underline{e}^T \theta_0^{-1} \underline{1}} \tag{7.3.26}$$

The ruin probabilities may also be obtained in the same manner, i.e., one finds

$$\underline{R}^T = \underline{\Theta}_0^T \theta_0^{-1} + T_1 \underline{e}^T \theta_0^{-1}, \tag{7.3.27}$$

with T_1 given by (7.3.26). Equation (7.3.25) allows calculation of T_2, giving

$$\left[\frac{\underline{e}^T \theta_0^{-1} \underline{1}}{2}\right] T_2 = \underline{\Theta}_1^T \theta_0^{-1} \underline{1} - T_1 - \underline{R}^T \theta_1 \theta_0^{-1} \underline{1}, \tag{7.3.28}$$

where T_1 and $\underline{R}^T$ are given by (7.3.26) and (7.3.27), respectively. The variance is, of course, $T_2 - T_1^2$.

§7.4. Covariance and Spectral Density Structure for Time-Reversible Processes.

Let $N(t)$ be any finite ergodic stationary time-reversible Markov chain. Its covariance function is given by

$$r_N(\tau) = E[N(t)\,N(t+\tau)] - E^2[N(t)]. \tag{7.4.1}$$

Hence

$$r_N(\tau) = \sum_m \sum_n m\, e_m\{p_{mn}(\tau) - e_n\}n. \tag{7.4.2}$$

From the spectral representation (3.2.6), we then have

$$r_N(\tau) = \sum_{j=2}^{N} e^{-\alpha_j|\tau|}\left(\sum_m m\, u_m^{(j)}\sqrt{e_m}\right)^2 = \sum_{j=2}^{N} \theta_j e^{-\alpha_j|\tau|}, \quad \theta_j \geq 0 \tag{7.4.3}$$

and this is completely monotone on $(0,\infty)$. The same reasoning clearly goes through for $X(t) = f(N(t))$ where f is any function defined on the state space. In summary we have

Theorem 7.4A. Let $N(t)$ be a finite ergodic stationary time-reversible Markov chain on state space $\mathcal{N}$, and let $f(n)$ be any function defined on $\mathcal{N}$. Then $r_f(\tau) = \mathrm{cov}[f(N(t))f(N(t+\tau))]$ is completely monotone in τ on $(0,\infty)$.

The covariance function, given on $(0,\infty)$ by

$$r_f(\tau) = \sum_m \sum_n f(m)\, e_m(p_{mn}(\tau) - e_n)\, f(n) \tag{7.4.4}$$

is such that

$$\int_0^\infty r_f(\tau)d\tau = \underline{f}^T\, \mathbf{e}_D Z\underline{f}, \tag{7.4.5}$$

where Z is the fundamental matrix of the chain discussed in §6.10. From Lemma 7.2A we see that $\mathbf{e}_D Z$ is positive semi-definite. From the proof of Theorem 7.2C we see that $\underline{f}^T\, \mathbf{e}_D\, Z\underline{f}$ is strictly positive unless $f(n)$ is independent of n. We note that (7.4.5) does not require reversibility.

Example 7.4B. Let $N(t)$ be the reliability system with independent Markov components as in 6.3A and 6.5A. Let $i_G(N(t))$ be the indicator function for the good set, with values $i_G(n) = 1$, $n \in G$; $i_G(n) = 0$, $n \in B$. This may be called the performance process for the system. The theorem states that $r_{i_G}(\tau)$ is completely monotone.

Two structural features of such completely monotone covariance functions should be noted.

Remark 7.4C. The spectral density for the process is always unimodal. Indeed, since the covariance function is a mixture of two sided exponentials, the spectral density function [9]

$$\begin{aligned} f_X(\lambda) &= \frac{1}{2\pi} \int_{-\infty}^{\infty} e^{-i\tau\lambda} \, r_X(\tau) d\tau \\ &= \frac{1}{\pi} \sum_{j=2}^{N} \frac{\sigma_j \alpha_j}{\lambda^2 + \alpha_j^2}. \end{aligned} \tag{7.4.6}$$

Remark 7.4D. For simple kinds of smoothing of the process $f(N(t))$, the covariance function is positive and unimodal, and the spectral density function is unimodal. For, let

$$Y(t) = \int_0^{\infty} A(t-t') \, X(t')dt', \tag{7.4.7}$$

where $A(t)$ is a non-negative smoothing function, $\zeta e^{-\zeta t}$ for example. Then if $A(t)$ is integrable, $Y(t)$ will have the covariance function

$$r_Y(\tau) = \int_0^{\infty} r_A(\tau-\tau') \, r_X(\tau')d\tau' \tag{7.4.8}$$

where

$$r_A(\tau) = \int_0^{\infty} A(t) \, A(t+\tau)dt. \tag{7.4.9}$$

When $A(t)$ is log-concave, as is true for most simple smoothing functions, then $r_A(\tau)$ is log-concave by Ibragimov's Theorem on the preservation of log-concavity under convolution

[19], and strongly unimodal. The positivity and unimodality of $r(\tau)$ then follow. For such $A(t)$, the spectral density is also unimodal, as the reader will verify.

Some comments on the nature of the covariance function of Example 7.4B for the performance process of the system may be of interest. If we call the covariance function $r_G(\tau)$ for brevity, we note that $r_G(0) = \overline{i_G^2} - (\bar{i}_G)^2 = \bar{i}_G - (\bar{i}_G)^2$ since $i_G^2 = i_G$. Hence

$$r_G(0) = P(G)[1-P(G)]. \tag{7.4.10}$$

where $P(G) = \sum_G e_n$ is the ergodic probability of the good set. If we look at $r_B(\tau)$, we find that, since $i_G + i_B = 1$,

$$r_G(\tau) = r_B(\tau). \tag{7.4.11}$$

If we consider the simple system with one Markov component, and failure rates λ, μ, the reader will quickly verify that

$$r_G(\tau) = r_B(\tau) = \frac{\lambda\mu}{(\lambda+\mu)^2} e^{-(\lambda+\mu)\tau}. \tag{7.4.12}$$

The correlation function $r^*(\tau) = r(\tau)/r(0)$ is therefore

$$r^*(\tau) = e^{-(\lambda+\mu)\tau}. \tag{7.4.13}$$

This corresponds to the "relaxation time" $T^* = (\lambda+\mu)^{-1}$ for this elementary system. It is the time constant for the rate of approach to ergodicity. We note that $T^* = (T_F^{-1} + T_R^{-1})^{-1}$ where T_F and T_R are the failure time and repair time. The heuristic relaxation time defined by

$$T^* = \int_0^\infty r^*(\tau)d\tau, \tag{7.4.14}$$

is of possible value because of the accessibility of $r^*(\tau)$

from real data. Some further observations are made in §8.10.

§7.5. A Central Limit Theorem.

The positivity of (7.4.5) has a natural meaning in that it corresponds to a variance in a central limit theorem related to the chain.

Consider the additive process

$$S(t) = \int_0^t f(N(t'))dt', \tag{7.5.1}$$

where $N(t)$ and $f(N(t))$ are as in Theorem 7.4A. Then as one might expect and as may be proven in a variety of ways, the process (7.5.1) is asymptotically normal in distribution for large t, and satisfies a central limit theorem [40]. This states that

$$\frac{S(t) - \mu t}{\sigma\sqrt{t}} \overset{d}{\to} N(0,1), \quad \text{as} \quad t \to \infty. \tag{7.5.2}$$

Consequently, $S(t)$ also satisfies a weak law of large numbers stating that

$$\frac{S(t)}{t} \overset{d}{\to} \mu, \quad \text{as} \quad t \to \infty. \tag{7.5.3}$$

Clearly

$$\mu = \sum_{\mathcal{N}} e_n \; f(n) = \underline{e}^T \underline{f} = E[f(N(\infty))].$$

The coefficient σ needed for the central limit theorem, may be obtained as follows. We want $\sigma^2 = \lim t^{-1} \operatorname{Var}[S(t)] = \lim t^{-1} \operatorname{Var}[S(t)-\mu t] = \lim t^{-1} E[(S(t)-\mu t)^2]$. But if $X(t) = f(N(t))-e^T f$, then

$$E(S(t)-\mu t)^2 = E\int_0^t X(t')\,dt' \int_0^t X(t'')dt''$$

$$= \int_0^t dt' \int_0^t dt'' \; r_f(t'-t'') = \int_0^t R(w)dw,$$

where $R(w) = 2\int_0^w r(u)du$. Since $r(u)$ is integrable, we have $R(w) \to 2\int_0^\infty r(u)du$, as $w \to \infty$. Hence from L'Hospitals rule, and (7.4.5)

$$\sigma^2 = 2\int_0^\infty r_f(\tau)d\tau = 2f^T e_D Z\underline{f}. \qquad (7.5.4)$$

The central limit theorem is useful in that it quantifies the element of risk in a particular system. For a reliability system, say, the allocation of expenditures to different components may require the minimization of total expenditure subject to a minimal expected duty cycle for the performance of the system. If down times are very costly, the designer must be risk averse, and will also impose a constraint on the maximal permissible σ.

§7.6. Regeneration Times and Passage Times - Their Relation For Arbitrary Chains.

In many ergodic processes of interest, there is a special state in the good set G, the returns to which provide a useful focal point for the study of passage to the bad set B. For birth-death processes this special state might be the extreme state $n = 0$. For the systems of independent Markov components of Examples 6.3A and 6.5A the special state might be the perfect state. The results of this section will be valid for any arbitrary finite ergodic chain $N(t)$.

Let A be the special state in set G. Let all the states of the bad set B be *aggregated* to a single state B by the modification technique of 3.5C which is valid for non-reversible chains as well as reversible chains. The modified process $N_M(t)$ will still be ergodic and the passage time density $s_{AB}(\tau)$ from A to set B for $N(t)$ will be identical to $s_{AB}(\tau)$ from state A to state B for $N_M(t)$. Some notation is needed.

Let $r_A(\tau)$ be the regeneration time density for the modified process $N_M(t)$, i.e., the density of time intervals between returns to state A.

Let $s_A(\tau)$ be the sojourn time density (§6.5) for $N_M(t)$ on the set $\mathcal{N}_M$-A, i.e., on the complement of A. Clearly

$$r_A(\tau) = \nu_A e^{-\nu_A \tau} * s_A(\tau). \tag{7.6.1}$$

Let θ_A^+ be the probability that the process $N_M(t)$ after leaving A will return to A without reaching state B.

Let $r_A^+(\tau)$ be the conditional regeneration time density conditioned on not reaching B.

Let $s_{AB}^+(\tau)$ be the conditional passage time density for $N_M(t)$, conditioned on not returning to A before getting to B.

Let $s_{BA}(\tau)$ be the passage time density for $N_M(t)$ from B to A.

We then have the following self-consistency relations

$$s_{AB}(\tau) = \theta_A^+ \, r_A^+(\tau) * s_{AB}(\tau) + (1-\theta_A^+) \, s_{AB}^+(\tau), \tag{7.6.2}$$

$$r_A(\tau) = \theta_A^+ \, r_A^+(\tau) + (1-\theta_A^+) \, s_{AB}^+(\tau) * s_{BA}(\tau). \tag{7.6.3}$$

Then from Laplace transformation

$$\sigma_{AB}(s) = \theta_A^+ \rho_A^+(s) \, \sigma_{AB}(s) + (1-\theta_A^+) \, \sigma_{AB}^+(s), \tag{7.6.4}$$

$$\rho_A(s) = \theta_A^+ \, \rho_A^+(s) + (1-\theta_A^+) \, \sigma_{AB}^+(s) \, \sigma_{BA}(s). \tag{7.6.5}$$

We now solve for $\theta_A^+ \rho_A^+(s)$ and $(1-\theta_A^+) \, \sigma_{AB}^+(s)$ and find that

$$(1-\theta_A^+)\ \sigma_{AB}^+(s) = \frac{\sigma_{AB}(s)[1 - \rho_A(s)]}{1 - \sigma_{AB}(s)\ \sigma_{BA}(s)}\ , \tag{7.6.6}$$

$$\theta_A^+\ \rho_A^+(s) = \frac{\rho_A(s) - \sigma_{AB}(s)\ \sigma_{BA}(s)}{1 - \sigma_{AB}(s)\ \sigma_{BA}(s)}\ . \tag{7.6.7}$$

If we now let $s \to 0+$ in either (7.6.6) or (7.6.7) we find from L'Hospitals rule

$$\theta_A^+ = 1 - \frac{\overline{T}_R(A)}{\overline{T}_{AB}+\overline{T}_{BA}}\ , \tag{7.6.8}$$

where $\overline{T}_R(A)$ is the mean regeneration time for the modified process, $\overline{T}_{AB}$ is the mean passage time from A to B, and $\overline{T}_{BA}$ is the mean passage time from B to A. The following theorem is now valid.

<u>Theorem 7.6A</u>. Let $N(t)$ be any finite ergodic chain, and let A, B be any pair of states in the state space $\mathscr{N}$. Then for the mean regeneration times and mean passage times we have

$$\overline{T}_R(A) = (\overline{T}_{AB}+\overline{T}_{BA})(1-\theta_A^+) \le \overline{T}_{AB}+\overline{T}_{BA}. \tag{7.6.9}$$

The proof is immediate from (7.6.8) since the partition of the state space is arbitrary. There is strict inequality if $\mathscr{N}$ has more than two states.

The reader will note that $\overline{T}_R(A)$ has a very simple form, i.e., in the context of Theorem 7.5A

$$\overline{T}_R(A) = \frac{1}{e_A \nu_A}\ , \tag{7.6.10}$$

where e_A is the ergodic probability of state A for $N(t)$. To see that this is so, we observe from (7.6.1) that

$$\overline{T}_R(A) = \frac{1}{\nu_A} + T_V(\overline{A}), \tag{7.6.11}$$

where $T_V(\overline{A})$ is the mean sojourn time on $\mathscr{N}-A = \overline{A}$. But from

(6.7.11)

$$T_V(\bar{A}) = \frac{1-e_A}{i_{A\bar{A}}} = \frac{1-e_A}{\nu_A e_A}, \tag{7.6.12}$$

and (7.6.10) follows. From (7.6.9) and (7.6.10) we then have

Corollary 7.6B. In the context of Theorem 7.6A

$$\bar{T}_{AB}+\bar{T}_{BA} = \frac{(\nu_A e_A)^{-1}}{1-\theta_A^+} = \frac{(\nu_B e_B)^{-1}}{1-\theta_B^+} . \tag{7.6.13}$$

We note from (7.6.13) that $\bar{T}_{AB}+\bar{T}_{BA} \geq \frac{1}{2}\max[(\nu_A e_A)^{-1},(\nu_B e_B)^{-1}]$ for every ergodic chain.

Remark 7.6C. Suppose we have a time-reversible chain and we want $\bar{T}_{AB}$, the mean time from state A to set B. If as in 3.5C we aggregate the states of B, $\bar{T}_{AB}$ is unchanged, and e_A, e_B, ν_A, and ν_B for the modified process are still at hand, then

$$\bar{T}_{AB}+\bar{T}_{BA} = \frac{1}{e_B\nu_B}\,\frac{1}{1-\theta_B^+} . \tag{7.6.14}$$

If it is known that e_B is small, $\theta_B^+ \ll 1$, and $\bar{T}_{BA} \ll \bar{T}_{AB}$, as will often be the case, then $\bar{T}_{AB}$ is closely approximated by $\frac{1}{e_B\nu_B} \approx \frac{1-e_B}{e_B\nu_B}$, i.e., $\bar{T}_{AB} \approx T_V(G)$ where $\mathcal{N} = G+B$. The quantity θ_B^+ is sensitive to jitter (§6.10), and may jeopardize the requirement that $\theta_B^+ \ll 1$.

§7.7. Passage to a Set with Two States.

Let $N(t)$ be any finite ergodic chain on a state space $\mathcal{N}$. Let $0 \in \mathcal{N}$, and let α and β be two other states of $\mathcal{N}$. Some simple aspects of the passage from 0 to $B^* = \{\alpha,\beta\}$ are of interest and are needed for the next chapter.

Equation (7.3.6) now has the form, with a slight change of notation

$$\pi_{0\alpha}(s) = \rho_{0\alpha}(s)\pi_{\alpha\alpha}(s) + \rho_{0\beta}(s)\ \pi_{\beta\alpha}(s), \tag{7.7.1a}$$

$$\pi_{0\beta}(s) = \rho_{0\alpha}(s)\pi_{\alpha\beta}(s) + \rho_{0\beta}(s)\ \pi_{\beta\beta}(s). \tag{7.7.1b}$$

If we divide (7.7.1a) by $\pi_{\alpha\alpha}(s)$ and (7.7.1b) by $\pi_{\beta\beta}(s)$ and make the identification,

$$\frac{\pi_{mn}(s)}{\pi_{nn}(s)} = \sigma_{mn}(s) = E[e^{-sT_{mn}}],$$

where T_{mn} is the first passage time from m to n, we have

$$\sigma_{0\alpha}(s) = \rho_{0\alpha}(s) + \rho_{0\beta}(s)\ \sigma_{\beta\alpha}(s), \tag{7.7.2a}$$

$$\sigma_{0\beta}(s) = \rho_{0\alpha}(s)\sigma_{\alpha\beta}(s) + \rho_{0\beta}(s). \tag{7.7.3b}$$

It follows that

$$\rho_{0\alpha}(s) = \frac{\sigma_{0\alpha}(s) - \sigma_{0\beta}(s)\ \sigma_{\beta\alpha}(s)}{1 - \sigma_{\alpha\beta}(s)\sigma_{\beta\alpha}(s)}, \tag{7.7.4a}$$

$$\rho_{0\beta}(s) = \frac{\sigma_{0\beta}(s) - \sigma_{0\alpha}(s)\ \sigma_{\alpha\beta}(s)}{1 - \sigma_{\alpha\beta}(s)\sigma_{\beta\alpha}(s)}. \tag{7.7.4b}$$

We see that the numerator and denominator on the right hand side of each of these equations vanishes at $s = 0$. L'Hospitals rule then gives for (7.7.4a)

$$0 \leq \rho_{0\alpha}(0) = \frac{-\overline{T}_{0\alpha} + (\overline{T}_{0\beta}+\overline{T}_{\beta\alpha})}{\overline{T}_{\alpha\beta} + \overline{T}_{\beta\alpha}} \leq 1 \tag{7.7.5}$$

since the term on the left is a ruin probability, i.e., the probability of reaching α from 0 before B. A similar result is obtained from (7.7.4b). The finiteness of the chain does not matter - only the ergodicity has been used. We then have

<u>Theorem 7.7A</u>. For any ergodic Markov chain $N(t)$ in continuous time, one has the "triangle inequality"

$$\bar{T}_{\alpha\gamma} \le \bar{T}_{\alpha\beta} + \bar{T}_{B\gamma}, \tag{7.7.6}$$

for every triplet of states α,β,γ in the state space. This theorem can be strengthened to deal with passage times to a set. The proof will be given in §8.5.

<u>Theorem 7.7B</u>. Let $N(t)$ be any ergodic Markov chain in continuous time, with state space $\mathscr{N}$. Let B be a proper subset of $\mathscr{N}$ and let m,n be states in $\mathscr{N}$-B. Then for the passage times T_{mB} and T_{nB} from m and n to set B one has the triangle inequality

$$\bar{T}_{mB} \le \bar{T}_{mn} + \bar{T}_{nB}. \tag{7.7.7}$$

This triangle inequality can be strengthened. The basic relation is one of stochastic ordering. The proof requires only the conditioning ideas present in (7.6.2) and (7.6.3), as pointed out by D. R. Smith.

<u>Theorem 7.7C</u>. Let $N(t)$ be any ergodic chain on state space $\mathscr{N}$, and let α,β,γ be any three (distinct) states of $\mathscr{N}$. Then

$$T_{\alpha\gamma} \prec T_{\alpha\beta} + T_{\beta\gamma} \tag{7.7.8}$$

<u>Proof</u>: A path going from α to γ either reaches γ before reaching β or reaches β first and then gets to γ subsequently. Let $P_{\alpha\gamma}$ = P[γ is reached from α before reaching β] and let $P_{\alpha\beta}$ = P[β is reached from α before γ]. Let $s_{\alpha\gamma|\beta}(\tau)$ be the p.d.f. of $T_{\alpha\gamma}$ given that β is reached before γ, and let $s_{\alpha\gamma|\gamma}$ be the p.d.f. of $T_{\alpha\gamma}$ given that γ is reached before β. Then, with asterisks denoting convolution,

$$s_{\alpha\gamma}(\tau) = [\overbrace{s_{\alpha\beta|\beta}(\tau) * s_{\beta\gamma}(\tau)}^{s_{\alpha\gamma|\beta}}]P_{\alpha\beta} + s_{\alpha\gamma|\gamma}(\tau)\, P_{\alpha\gamma}. \tag{7.7.9a}$$

Similarly, interchanging β and γ

$$s_{\alpha\beta}(\tau) = [s_{\alpha\beta|\beta}(\tau)]P_{\alpha\beta} + [s_{\alpha\gamma|\gamma}(\tau) * s_{\gamma\beta}(\tau)]P_{\alpha\gamma}. \quad (7.7.9b)$$

From (7.7.9b)

$$s_{\alpha\beta}(\tau) * s_{\beta\gamma}(\tau) = [s_{\alpha\beta|\beta}(\tau) * s_{\beta\gamma}(\tau)]P_{\alpha\beta} + [s_{\alpha\gamma|\gamma}(\tau) * s_{\gamma\beta}(\tau) * s_{\beta\gamma}(\tau)]P_{\alpha\gamma}. \quad (7.7.10)$$

A comparison of (7.7.10) with (7.7.9a) shows that, since $T_{\alpha\gamma|\gamma} + T_{\gamma\beta} + T_{\beta\gamma} \succ T_{\alpha\gamma|\gamma} + T_{\gamma\beta}$, and $s_3(\tau) \succ s_2(\tau)$ implies that $p_1s_1(\tau) + p_2s_3(\tau) \succ p_1s_1(\tau) + p_2s_2(\tau)$, one has (7.7.8).

Note that Theorem 7.7A is then a corollary of Theorem 7.7C. Indeed $\varepsilon[T^k_{\alpha\gamma}] \le \varepsilon[(T_{\alpha\beta}+T_{\beta\gamma})^k]$ for $k \ge 0$, when the moments exist.

The reasoning of Theorem 7.7B permits, by the same argument, after aggregation.

<u>Theorem 7.7D</u>. In the setting of Theorem 7.7B one has

$$T_{mB} \prec T_{mn} + T_{nB}.$$

By the identical reasoning, one has also

<u>Theorem 7.7E</u>. In the setting of Theorem 7.7C, let $T_{\alpha\alpha}$ be the regeneration time for state $\alpha \in \mathcal{N}$. Then for any distinct $\alpha,\beta \in \mathcal{N}$, one has

$$T_{\alpha\alpha} < T_{\alpha\beta} + T_{\beta\alpha}.$$

The reader will note that equations (7.7.2a,b) are valid even if the chain $N(t)$ is lossy or transient. Consequently, from the positivity of the numerators in (7.7.4a,b) at $s = 0$, one has:

<u>Theorem 7.7F</u>. Let $N(t)$ be any lossy or transient chain, with states α,β,γ in its state space. Let R_{mn} be the

probability of ever reaching state n from state m before leaving the state space. Then

$$R_{\alpha\gamma} > R_{\alpha\beta} R_{\beta\gamma}. \qquad (7.7.8)$$

<u>Example 7.7G</u>. Consider a reliability system modeled by a chain $N(t)$ on $\mathscr{N} = G+B$, with B representing failure states. Let G^* be the "critical" states, i.e., the subset of G from which transfers to state B can occur, and let $g,h,i,\ldots,$ be the states of G^*. Then R_{gh} is the probability of reaching h from g before failing.

Chapter 8

Rarity and Exponentiality

§8.0. Introduction.

If a system is modeled by a finite Markov chain which is ergodic, the passage time from some specified initial distribution over the state space to a subset B of the state space visited infrequently is often exponentially distributed to good approximation. The chapter is devoted to the limit theorems surrounding such behavior for processes and the characterization of the circumstances under which exponentiality is present. In the absence of certain "jitter," i.e., clustering of the entry epochs into the good set G, the time to failure from the perfect state, the quasi-stationary exit time, the ergodic exit time and the sojourn time on the good set then have a common asymptotic exponential distribution and common expectations. For engineering purposes, it is essential to quantify departure from exponentiality via error bounds. When one is dealing with time-reversible chains e.g., systems with independent Markov components, the complete monotonicity present permits such quantification and the error bounds needed.

§8.1. Passage Time Density Structure for Finite Ergodic Chains; the Exponential Approximation.

The conditional regeneration time density $r_A^+(\tau)$ (cf. §7.6) for T_R^+ and direct passage time density $s_{AB}^+(\tau)$ are key components of the passage time density $s_{AB}(\tau)$ from state A to set B. In the notation of §7.6, (7.6.4) gives

$$\sigma_{AB}(s) = \left\{\frac{1 - \theta_A^+}{1 - \theta_A^+ \rho_A^+(s)}\right\} \sigma_{AB}^+(s), \tag{8.1.1}$$

with the real time analogue

$$s_{AB}(\tau) = \left\{\sum_0^\infty (1 - \theta_A^+)\theta_A^{+k}\, r_A^{+(k)}(\tau)\right\} * s_{AB}^+(\tau). \tag{8.1.2}$$

This has a simple probabilistic meaning. There will be a sequence of identically distributed cycles each consisting of a dwell time in A, and a sojourn in the state space $\mathcal{N}$-{A} which does not visit B. Finally the sequence will terminate with a different subcycle for which the sojourn does visit B without having returned to A. The non-visiting cycles have density

$$r_A^+(\tau) = \nu e^{-\nu\tau} * s_V^+(\tau), \tag{8.1.3}$$

where $s_V^+(\tau)$ is the sojourn time density on $\mathcal{N}$- {A} when B is not visited. The structure of (8.1.2) may be thought of as a sequence of Bernoulli trials with failure if B is visited, with probability of failure $1 - \theta_A^+$ at each trial. When visits to B are rare, this probability is small. The random number of successful cycles is geometrically distributed with

$$p_k = P[\text{k successes before failure}] = (1-\theta_A^+)\theta_A^{+k}, \tag{8.1.4}$$

and the expected number of successes is $\theta_A^+(1-\theta_A^+)^{-1}$. Correspondingly, from (8.1.2)

$$\overline{T}_{AB} = \frac{\theta_A^+}{1-\theta_A^+}\,\overline{T_R^+}(A) + \overline{T}_{AB}^+. \tag{8.1.5a}$$

When $(1-\theta_A^+) \ll 1$, the number of successes before failure is large. Since, moreover,

$$\overline{T_R^+(A)} = \int_0^\infty \tau r_A^+(\tau)\,d\tau = -\rho_A^{+\prime}(0) < \infty \tag{8.1.5b}$$

for $r_A^+(\tau)$ fixed the term in curly brackets in (8.1.2) converges in distribution (after rescaling) to the exponential distribution with mean one as $\theta_A^+ \to 1$. Specifically we have:

<u>Theorem 8.1A</u>. Let $(X_j)_1^\infty$ be a sequence of non-negative i.i.d. r.v.'s with c.d.f. $F_X(x)$ and positive finite expectation $\mu = E[X]$. Let K_θ be a geometrically distributed r.v. with $P[K_\theta = k] = (1-\theta)\theta^k$, independent of all X_j. Then

$$\frac{\sum_0^{K_\theta} X_j}{\left(\frac{\theta}{1-\theta}\right)\mu} \xrightarrow{d} Y \quad \text{as} \quad \theta \to 1-,$$

where Y has $F_Y(x) = 1 - e^{-x}$.

<u>Proof</u>: Let $Y_\theta = \sum_0^{K_\theta} X_j$, and $Z_\theta = Y_\theta/\{\theta(1-\theta)^{-1}\mu\} = Y_\theta/\mu_\theta$. Then $E[e^{-sY_\theta}] = (1-\theta)/\{1-\theta\phi(s)\}$ where $\phi(s) = E[e^{-sX}]$. Let $E[e^{-sY_\theta}] = (1-\theta)/[1-\theta\,\{1-\mu s + s\zeta(s)\}]$. It is easy to see that $\zeta(s) \to 0$ as $s \to 0+$. For

$$|\zeta(s)| = \left|\frac{\phi(s) - 1 + \mu s}{s}\right| = \left|\int_0^\infty \frac{e^{-sx} - 1 + sx}{sx}\,x\,dF_X(x)\right|$$

$$\le \int_0^\infty \left|\frac{e^{-sx} - 1 + sx}{sx}\right| x\,dF_X(x) \le M\int_0^\infty x\,dF_X(x) = M\mu.$$

We may then use the dominated convergence theorem to pass to the limit as $s \to 0+$ under the integral sign and find $\zeta(0+) = 0$. Then

$$E[e^{-sZ_\theta}] = E[e^{-(s/\mu_\theta)Y_\theta}] \to (1+s)^{-1}, \quad \text{as} \quad \theta \to 1-$$

and the convergence thoerem for characteristic functions (Feller II, [13], p. 408) may be employed.

<u>Remark 8.1B</u>. It may be expected that for a finite ergodic chain the passage time density (8.1.2) to a rare set with $(1 - \theta_A^+) << 1$ will be approximately exponential. To use the exponential approximation with conviction, one requires the following:

a) One must know that $(1 - \theta_A^+) << 1$.

b) One needs a good approximation to the mean time from A to set B, this being the single parameter needed for the exponential approximation.

c) One needs to know that $s_{AB}^+(\tau)$ the direct passage time density from A to B after the final visit to A will not disturb the exponentiality. One must know in practice that $E[T_{AB}^+] << \frac{\theta_A^+}{1-\theta_A^+} E[T_R^+(A)]$ in the notation of §7.7.

d) One needs to know how exponential $s_{AB}(\tau)$ is, i.e., one must have some meaningful measure of exponentiality that can be calculated for the system at hand. A useful error bound would be best.

§8.2. A Limit Theorem for Ergodic Regenerative Processes.

The essence of the exponential approximation takes its sharpest form in the context of a limit theorem for ergodic regenerative processes [26]. A process $N(t)$ is said to be ergodic and regenerative if there is at least one state in the state space which is regenerative, i.e., returns to this state are positive-recurrent and have an associated ergodic renewal process. The process loses its memory at these return states. All ergodic Markov chains in continuous time are regenerative. Indeed, all of their states are regenerative.

Consider the following setting.

Context 8.2A.

a) The state space $\mathscr{N}$ is not finite nor necessarily discrete. There is an infinite sequence of nested partitions available of the form $\mathscr{N} = G_K + B_K$, with $G_K \subset G_{K+1}$, $G_K \to \mathscr{N}$.

b) The state $0 \in G_1$ and is regenerative.

c) $E[T_{0B_K}] \to \infty$, as $K \to \infty$, where T_{0B_K} is the passage time from state 0 to set B_K.

As an example, $N(t)$ might be a birth-death process governed by $\{\lambda_n, \mu_n\}$ with $\lambda_n = \lambda$, $\mu_n = n\mu$ for $n \geq 0$. For $\mathscr{N} = \{0,1,2,\ldots\}$, one might have the partition $G_N = \{0,1,2,\ldots,N\}$ and $n = 0$ might be the regenerative state of central interest. T_{0B_N} is the time from 0 to $N + 1$. Here, clearly, $E[T_{0B_N}] \to \infty$ as $N \to \infty$.

Theorem 8.2B. In Context 8.2A,

$$\frac{T_{0B_K}}{E[T_{0B_K}]} \xrightarrow{d} Y, \tag{8.2.1}$$

where $F_Y(x) = 1 - e^{-x}$. Moreover,

$$(1 - p_K)\frac{E\,T_{0B_K}}{ET_R} \to 1, \tag{8.2.2}$$

where p_K is the probability of regenerating at state 0 before hitting set B_K and T_R is the regeneration time density for the process $N(t)$.

Proof: Let the process $N(t)$ start at $t = 0$ with the zero-th regeneration. Let q_K be the probability that B_K is reached between two successive regenerations and let $p_K = 1 - q_K$. Let M_K denote the number of regenerations before the first visit to B_K and let $T_1, T_2, T_3, \ldots$ be the time intervals between successive regenerations. (Note that the T_i's are i.i.d.) Then

$$\begin{aligned} 0 \le T_{0B_K} \le T_1 \quad &, \quad M_K = 0; \\ \sum_1^{M_K} T_m \le T_{0B_K} \le \sum_1^{M_K+1} T_m, &\quad M_K > 0. \end{aligned} \tag{8.2.3}$$

Moreover,

$$P[M_K = m] = p_K^m q_K. \tag{8.2.4}$$

Let $F_{1K}(x)$ be the conditional c.d.f. of T_1, given that B_K was not reached and let $F_{2K}(x)$ be that of T_1, given that B_K was reached. If $F(x)$ is the unconditional distribution of T_1, then clearly

$$F(x) = p_K F_{1K}(x) + q_K F_{2K}(x). \tag{8.2.5}$$

For the subset of paths for which $M_K = m$ the time intervals $T_1, T_2, \ldots, T_m$ are independent with $T_1, \ldots, T_m$ having the distribution F_{1K} and T_{m+1} having the distribution F_{2K}. For this subset of paths, $T_1 + \ldots + T_m$ has the characteristic function $\phi_{1K}^m(s)$ and $T_1 + \ldots + T_{m+1}$ has the ch.f. $\phi_{1K}^m(s)\ \phi_{2K}(s)$, where $\phi_{iK}(s) = \int_0^\infty e^{-st}\ dF_{iK}(t)$, $i = 1,2$. It then follows from (8.2.3) and (8.2.4) by summation over m that, for $s > 0$,

$$\frac{q_K}{1 - p_K\phi_{1K}(s)} \geq E\left[e^{-sT_{0B_K}}\right] \geq \frac{q_K\phi_{2K}(s)}{1 - p_K\phi_{1K}(s)} \tag{8.2.6}$$

and that

$$\frac{p_K}{q_K}\mu_{1K} \leq E\left[T_{0B_K}\right] \leq \frac{p_K}{q_K}\mu_{1K} + \mu_{2K}, \tag{8.2.7}$$

where

$$\mu_{iK} = \int_0^\infty x dF_{iK}(x), \quad i = 1,2.$$

We note from (8.2.5) that $p_K\mu_{1K} + q_K\mu_{2K} = E[T_1] \stackrel{d}{=} \mu$, and from (8.2.7) that for any K, $q_K E[T_{0B_K}] \leq \mu$. It follows that (8.2.7) implies (8.2.2) provided that $q_K\mu_{2K} \to 0$ and that it follows that $q_K \to 0 \Leftrightarrow E[T_{0B_K}] \to \infty$. Moreover, it follows from (8.2.5) that $p_K\phi_{1K}(s) + q_K\phi_{2K}(s) = \phi(s)$ where $\phi(s) = E[e^{-sT_1}]$. Hence

$$\frac{1 - p_K\phi_{1K}(q_Ks)}{q_K} = \frac{1 - \phi(q_Ks)}{q_K} + \phi_{2K}(q_Ks). \tag{8.2.8}$$

When $q_K \to 0$, the first term on the right goes to μs when $K \to \infty$, by virtue of the finiteness of $E[T_1]$. If it were further true that

$$\phi_{2K}(q_Ks) \to 1, \quad K \to \infty, \tag{8.2.9}$$

it would follow from (8.2.6) and (8.2.8) that

$$\lim_{N\to\infty} E\left[e^{-sq_KT_{0B_K}}\right] = \frac{1}{1 + \mu s} \tag{8.2.10}$$

for all $s > 0$, and the theorem would follow at once from a variant of the continuity theorem for characteristic functions given, for example, in Feller II, [15], p. 408. It only remains to prove (8.2.9) and

$$q_K\mu_{2K} \to 0, \quad K \to \infty. \tag{8.2.11}$$

We note that (8.2.9) follows from (8.2.11), since

$$0 \leq \frac{1 - \phi_{2K}(q_K s)}{q_K} = \int_0^\infty \frac{1 - e^{-q_K sx}}{q_K sx} sxdF_{2K}(x)$$

$$\leq \int_0^\infty sxdF_{2K}(x) = s\mu_{2K}.$$

To prove (8.2.11), let $\varepsilon > 0$ be given, and choose $x_0 > 0$ such that $\int_{x_0}^\infty xdF(x) < \varepsilon$. Since $dF(x) - q_K dF_{2K}(x) \geq 0$, by (8.2.5) we have $\int_{x_0}^\infty xq_K dF_{2K}(x) < \varepsilon$. Moreover, $\int_0^{x_0} xq_K dF_{2K}(x) \leq x_0 q_K$. Consequently, $q_K\mu_{2K} = \int_0^\infty xq_K dF_{2K}(x) < \varepsilon + x_0 q_K < 2\varepsilon$ as soon as $q_K < \varepsilon/x_0$, proving (8.2.11).

§8.3. Prototype Behavior: Birth-Death Processes; Strongly Stable Systems.

At the end of §8.1, a set of criteria for the validity of the exponential approximation and requirements for its use was set down. A study of the passage times for an ergodic birth-death process will illustrate these criteria [27].

Theorem 8.3A. Let $N(t)$ be an ergodic birth-death process governed by $\{\lambda_n, \mu_n\}$, with state space $\mathcal{N} = \{0,1,2,\ldots\}$. Let $\mathcal{N} = G_N + B_N$ be a sequence of partitions with $G_N = \{0,1,2,\ldots,N\}$. If $\lambda_n/\mu_n \to \rho < 1$ as $n \to \infty$, then $E[T_{N,N+1}]/E[T_{0,N+1}] \to 1 - \rho$. To prove the theorem a lemma is needed.

Lemma 8.3B. If the sequence of positive numbers θ_n converges to θ with $0 \leq \theta < 1$, and $Q_n = \theta_n + \theta_n\theta_{n-1} + \ldots + \theta_n\theta_{n-1}\ldots\theta_2\theta_1$, then the sequence Q_n converges to $\frac{\theta}{1 - \theta}$.

As proof, suppose first that $\theta > 0$. We note that $Q_n = \theta_n(1 + Q_{n-1})$. Since $\theta_n \to \theta$, for any positive α sufficiently small we may find an N_α such that

$0 < \theta - \alpha < \theta_n < \theta + \alpha < 1$ for $n > N_\alpha$. If we define Q_n^* by $Q_n^* = Q_n$ for $n \le N_\alpha$, and $Q_n^* = (\alpha + \theta)(1 + Q_{n-1}^*)$ for $n > N_\alpha$ it follows by induction that $Q_n \le Q_n^*$ for $n \ge N_\alpha$. By the contraction mapping theorem Q_n^* has the limit $\frac{\theta + \alpha}{1 - (\theta+\alpha)}$. Using the same procedure with $\theta - \alpha$ and associated Q_n^{**}, we find that $Q_n^{**} \ge Q_n$, where Q_n^{**} has the limit $\frac{\theta - \alpha}{1 - (\theta-\alpha)}$. Since α may be as small as we wish, Q_n converges to $\frac{\theta}{1 - \theta}$. The case $\theta = 0$ follows from $0 \le Q_n \le Q_n^*$.

We may now prove the theorem. From (5.2.2) we have

$$\frac{\overline{T}_{n-1}^+}{\overline{T}_n^+} = \frac{\lambda_n}{\mu_n} - \frac{1}{\mu_n \overline{T}_n^+}, \quad n \ge 1. \tag{8.3.1}$$

We next show that the last term in (8.3.1) vanishes as $n \to \infty$. From $\overline{T}_0^+ = \lambda_0^{-1}$ we may write:

$$\frac{1}{\mu_n \overline{T}_n^+} = \frac{\lambda_0}{\mu_n} \frac{\overline{T}_{n-1}^+ \, \overline{T}_{n-2}^+ \cdots \overline{T}_0^+}{\overline{T}_n^+ \, \overline{T}_{n-1}^+ \cdots \overline{T}_1^+}. \tag{8.3.2}$$

From (8.3.1) we see that $\overline{T}_{n-1}^+/\overline{T}_n^+ < \lambda_n/\mu_n$. It then follows from (8.3.2) that

$$\frac{1}{\mu_n \overline{T}_n^+} < \frac{\lambda_0}{\mu_n} \frac{\lambda_n}{\mu_n} \frac{\lambda_{n-1}}{\mu_{n-1}} \cdots \frac{\lambda_1}{\mu_1} = \frac{\lambda_n}{\mu_n}\left\{\frac{\lambda_0}{\mu_1} \frac{\lambda_1}{\mu_2} \cdots \frac{\lambda_{n-1}}{\mu_n}\right\}. \tag{8.3.3}$$

The term in curly brackets is recognized to be the potential coefficient π_n for the birth-death process (§3.3) and hence is equal to Ke_n where e_n is the ergodic probability of state n and K is some positive constant. Since $e_n \to 0$, and $\lambda_n/\mu_n \to \rho$ as $n \to \infty$, $(\mu_n \overline{T}_n^+)^{-1} \to 0$, and from (8.3.1)

$$\frac{\overline{T}_{n-1}^+}{\overline{T}_n^+} \to \rho, \quad n \to \infty. \tag{8.3.4}$$

The theorem now follows quickly from the lemma. For

$$\frac{\sum_0^N \overline{T}_j^+}{\overline{T}_N^+} = 1 + \sum_0^{N-1} \frac{\overline{T}_j^+}{\overline{T}_N^+} = 1 + \frac{\overline{T}_{N-1}^+}{\overline{T}_N^+} + \frac{\overline{T}_{N-1}^+}{\overline{T}_N^+}\,\frac{\overline{T}_{N-2}^+}{\overline{T}_{N-1}^+} + \cdots$$

$$\cdots + \left[\frac{\overline{T}_{N-1}^+}{\overline{T}_N^+}\,\frac{\overline{T}_{N-2}^+}{\overline{T}_{N-1}^+} \cdots \frac{\overline{T}_0^+}{\overline{T}_1^+}\right].$$

From the lemma with $\theta_n = \overline{T}_{n-1}^+/\overline{T}_n^+$, and $\theta = \rho$ from (8.3.4) we have

$$\frac{\overline{T}_{0,N+1}}{\overline{T}_{N,N+1}} \to 1 + \frac{\rho}{1-\rho} = \frac{1}{1-\rho}. \qquad (8.3.6)$$

This completes the proof.

The theorem may be extended easily to interrelate $\overline{T}_{0,N+1}$, $\overline{T}_{Q\ N+1}$, $\overline{T}_{E\ N+1}$, and $\overline{T}_{V\ N+1}$, these being the time from the bottom state to the set B_N, and the ergodic exit time, the quasi-stationary exit time and ergodic sojourn time on the set $G_N = \{0,1,2,\ldots,N\}$, respectively of Chapter 6. We note that $\overline{T}_{VN} = \overline{T}_{N,N+1}$.

Theorem 8.3C. Under the conditions of Theorem 8.3A, the mean exit times $\overline{T}_{0,N+1}$, $\overline{T}_{Q\ N+1}$ and $\overline{T}_{E\ N+1}$, are asymptotically equal, i.e., the ratio of any pair of these has the limit one as $N \to \infty$. Moreover,

$$\lim_{N\to\infty} \frac{\overline{T}_{0,N+1}}{\overline{T}_{VN}} = \lim_{N\to\infty} \frac{\overline{T}_{Q\ N+1}}{\overline{T}_{VN}} = \lim_{N\to\infty} \frac{\overline{T}_{E\ N+1}}{\overline{T}_{VN}} = \frac{1}{1-\rho} \qquad (8.3.7)$$

and

$$\overline{T}_{0,N+1} > \overline{T}_{Q\ N} > \overline{T}_{E\ N} > \overline{T}_{V\ N} \quad \text{for all } N. \qquad (8.3.8)$$

Proof: The demonstration of (8.3.7) is obtained from

$$\lim_{N\to\infty} \frac{\overline{T}_{m,N+1}}{\overline{T}_{N,N+1}} = \frac{1}{1-\rho}, \quad 0 \le m < \infty, \qquad (8.3.9)$$

which follows from $\overline{T}_{m,N+1} = \overline{T}_{0,N+1} - \overline{T}_{0m}$ and Theorem 8.3A. Moreover,

$$\zeta_{m,N} = \frac{\overline{T}_{m,N+1}}{\overline{T}_{N,N+1}} = \frac{\overline{T}_{0,N+1}}{\overline{T}_{N,N+1}} - \frac{\overline{T}_{0m}}{\overline{T}_{N,N+1}} < K \qquad (8.3.10)$$

for some positive K and all m,N since $\zeta_N = \overline{T}_{0,N+1}/\overline{T}_{N,N+1}$ converges, as $N \to \infty$. Consider the ratio

$$\frac{\overline{T}_{EN}}{\overline{T}_{VN}} = \frac{\sum_0^N e_n \zeta_{n,N}}{E_N} = \frac{\sum_0^\infty e_n U_{n,N}}{E_N}, \qquad (8.3.11)$$

where $E_N = \sum_0^N e_n$, and $U_{n,N} = \zeta_{n,N}$, $n \le N$, $U_{n,N} = 0$, $n > N$. Clearly $U_{n,N} < K$, all n, N. Hence we may use the dominated convergence theorem for the numerator of the right hand term when we let $N \to \infty$. From (8.3.9) we then have

$$\lim_{N\to\infty} \frac{\overline{T}_{EN}}{\overline{T}_{VN}} = \frac{1}{1-\rho}. \qquad (8.3.12)$$

We have seen previously (§6.9) that the four exit times of interest have the stochastic order

$$T_{0,N+1} \succ T_{QN} \succ T_{EN} \succ T_{VN} \qquad (8.3.13)$$

implying (8.3.8). Then $\overline{T}_{QN}/\overline{T}_{VN}$ is caught between $(\overline{T}_{0,N+1}/\overline{T}_{VN})$ and $(\overline{T}_{EN}/\overline{T}_{VN})$ each of which has the limit $(1-\rho)^{-1}$. The theorem then follows.

We have seen, (Eq. (5.2.10)), that $R_N^{(N+1)}$, the probability that a process starting at N reaches $N+1$ before reaching state zero is

$$R_N^{(N+1)} = \frac{\rho_N + (\rho_N\rho_{N-1}) + \dots + (\rho_N\rho_{N-1}\cdots\rho_1)}{1 + \rho_N + (\rho_N\rho_{N-1}) + \dots + (\rho_N\rho_{N-1}\cdots\rho_1)},$$

where $\rho_n = (\lambda_n/\mu_n)$. This has the form $Q_N/(1+Q_N)$ where Q_N is the expression of Lemma 8.3B with $\theta_n = \rho_n$. By that lemma Q_n converges to $\rho/(1-\rho)$ and $Q_n/(1+Q_n)$ converges to ρ. Hence

$$\lim_{N\to\infty} R_N^{(N+1)} = \rho. \tag{8.3.14}$$

The parameter ρ is therefore the asymptotic probability for large N of returning to $N+1$ after reaching the set G_N (from $N+1$) before reaching state zero.

Remark 8.3D. Note that $\bar{T}_{VN}$, the mean sojourn time on G_N, is given by (6.7.11)

$$\bar{T}_{VN} = \frac{\sum_0^N e_n}{\lambda_N e_N}. \tag{8.3.15}$$

Theorem 8.3C implies that $\bar{T}_{0,N+1}$, $\bar{T}_{QN}$, and $\bar{T}_{EN}$ may all be approximated for large N via the mean sojourn time (8.3.15) available from the ergodic distribution and the transition rates λ_m, μ_m. From (8.3.15), as $N\to\infty$ $\bar{T}_{VN} \sim (\lambda_N e_N)^{-1} = (\mu_{N+1}e_{N+1})^{-1}$. From (5.2.3) we see that $\mu_{N+1}e_{N+1}$ behaves as $K\rho^N$ for large N, so that all the exit times are becoming infinite. Correspondingly from any fixed state m the probability of not reaching the state 0 before the state N vanishes as $N\to\infty$. This may also be seen from (5.2.10). The limit theorem of §8.2 is therefore applicable for this process, and the exponential approximation

$$S_{0N}(\tau) \approx \frac{1}{\mu_{VN}} e^{-\tau/\mu_{VN}}$$

improves with N.

Remark 8.3E. The asymptotic ratio $(1-\rho)^{-1}$ in Theorem 8.3C has a simple interpretation. The sojourns on $G_N = \{0,1,2, \ldots, N\}$ commence at N. With probability θ_N^+, the sojourn will end by return to $N+1$ before reaching state

0. With probability $1 - \theta_N^+$, the sojourn will visit state 0 and then return to state $N + 1$. When N is large there will be many visits to state 0 before the end of the sojourn. Hence (cf. (8.1.2) and §7.5),

$$s_{N,N+1}(\tau) = \theta_N^+ s_{N,N+1}^+(\tau) + (1 - \theta_N^+) s_{0,N+1}(\tau) * s_{N,0}^+(\tau).$$

One has $\overline{T}_{N,N+1}^+/\overline{T}_{0N} \to 0$ as $N \to \infty$, since the denominator is becoming infinite. Consequently,

$$\overline{T}_{VN} = \overline{T}_{N,N+1} \sim (1 - \theta_N^+)\overline{T}_{0,N+1}, \quad \text{as} \quad N \to \infty$$

and the parameter ρ in (8.3.7) has been identified as the limit of the ruin probability θ_N^+. The character of a typical sample path is shown below. The other mean exit times

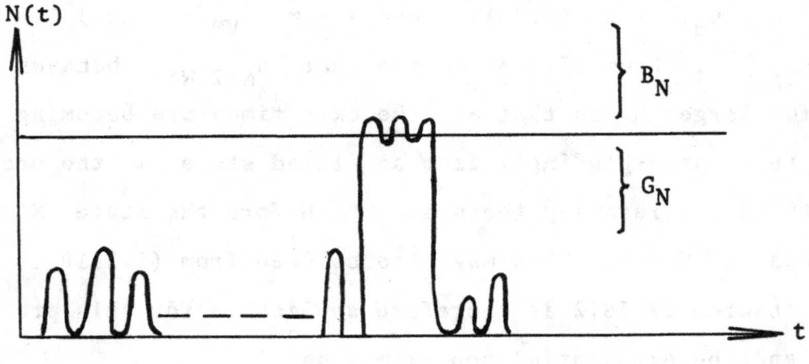

are asymptotically equal to $\overline{T}_{0,N+1}$ as $N \to \infty$ because the probability of hitting $N + 1$ before hitting 0 from state m with m fixed goes to zero as $N \to \infty$, and the mean time $\overline{T}_{m0}$ is asymptotically negligible compared to $\overline{T}_{0,N+1}$.

§8.4. Limiting Behavior of the Ergodic and Quasi-stationary Exit Time Densities and Sojourn Time Densities for Birth-Death Processes.

Consider the ergodic exit time density $s_E(\tau)$ for an ergodic birth-death process $N(t)$ for which $\overline{T}_{ON} \to \infty$, so that from §8.2,

$$\frac{T_{0,N}}{\overline{T}_{0,N}} \xrightarrow{d} Y. \tag{8.4.1}$$

Then for $\sigma_{mN}(s) = E[e^{-sT_{mN}}]$, we have

$$\sigma_{ON}\left(\frac{s}{\overline{T}_{0,N+1}}\right) = \sigma_{Om}\left(\frac{s}{\overline{T}_{0,N+1}}\right)\,\sigma_{mN}\left(\frac{s}{\overline{T}_{0,N+1}}\right). \tag{8.4.2}$$

When m is held fixed and $N \to \infty$, $\sigma_{Om}(s/\overline{T}_{ON}) \to 1$ by virtue of the continuity of $\sigma_{Om}(s)$ at $s = 0$. Hence

$$\lim_{N\to\infty} \sigma_{mN}\left(\frac{s}{\overline{T}_{0,N+1}}\right) = \lim_{N\to\infty} \sigma_{ON}\left(\frac{s}{\overline{T}_{0,N+1}}\right) = \frac{1}{1+s} \tag{8.4.3}$$

from Theorem 8.2B. It follows that the ergodic exit time density $s_{EN}(\tau)$ for the set $G_N = \{0,1,2, \ldots, N\}$ is such that

$$\lim_{N\to\infty} \sigma_{EN}\left(\frac{s}{\overline{T}_{0,N+1}}\right) = \lim_{N\to\infty} \frac{\sum_0^N e_m \sigma_{mN}\left(\frac{s}{\overline{T}_{0,N+1}}\right)}{\sum_0^N e_m}. \tag{8.4.5}$$

Hence by the dominated convergence theorem, since $\sigma_{mN}(s) < 1$, $s \geq 0$, one has

$$\lim_{N\to\infty} \sigma_{EN}\left(\frac{s}{\overline{T}_{ON+1}}\right) = \frac{1}{1+s}. \tag{8.4.6}$$

Consequently, one has:

Theorem 8.4A. For any ergodic birth-death process $N(t)$ for which $\overline{T}_{0N} \to \infty$, as $N \to \infty$, one has for $\mathcal{N} = G_N + B_N$, $G_N = \{0,1,2, \ldots, N\}$, and T_{EN} the ergodic exit time on G_N,

$$\frac{T_{EN}}{\overline{T}_{0,N+1}} \overset{d}{\to} Y \tag{8.4.7}$$

where Y is the exponential distribution with $F_Y(x) = 1-e^{-x}$.

Corollary 8.4B. In the same context, for the quasi-stationary exit time T_{QN}, one has

$$\frac{T_{QN}}{\overline{T}_{0,N+1}} \overset{d}{\to} Y. \tag{8.4.8}$$

Proof: From the stochastic ordering $T_0 \succ T_Q \succ T_E$ of Example 6.9D, one has since $X \succ Y \Rightarrow Ee^{-sX} \le Ee^{-sY}$, when $s \ge 0$,

$$\sigma_{EN}(s) \ge \sigma_{QN}(s) \ge \sigma_{0,N+1}(s), \quad s \ge 0. \tag{8.4.9}$$

Hence, setting $s = s^*/\overline{T}_{0,N+1}$, letting $N \to \infty$, and using Theorem 8.4A, and Theorem 8.2B, one finds (8.4.8). □

We next consider the ergodic sojourn time density $s_{VN}(\tau)$ on $G_N = \{0,1,2, \ldots, N\}$, and $\sigma_{VN}(s) = E[e^{-sT_{VN}}]$. From (6.7.1), $\sigma_{EN}(s) = \{1 - \sigma_{VN}(s)\}/s\overline{T}_{VN}$ and

$$\sigma_{EN}\left(\frac{s}{\overline{T}_{0,N+1}}\right) = \frac{1 - \sigma_{VN}\left(\frac{s}{\overline{T}_{0,N+1}}\right)}{\frac{\overline{T}_{VN}}{\overline{T}_{0,N+1}}\, s} . \tag{8.4.10}$$

Hence from (8.4.7), if $\left(\frac{\overline{T}_{VN}}{\overline{T}_{0,N+1}}\right) \to 1 - \rho$ as in Theorem 8.3C, we have

$$\lim_{N\to\infty} \sigma_{VN}\left(\frac{s}{\overline{T}_{0,N+1}}\right) = \rho + (1 - \rho)\left(\frac{1}{1 + s}\right) . \tag{8.4.11}$$

This gives:

Theorem 8.4C. For the ergodic birth-death process $N(t)$ for which $\frac{\lambda_n}{\mu_n} \to \rho$ of Theorem 8.3C

$$\frac{T_{VN}}{\overline{T}_{0,N+1}} \xrightarrow{d} X,$$

where $\overline{F}_X(x) = P[X > x] = (1 - \rho)e^{-x}$.

The term ρ in (8.4.11) corresponds to the contribution of jitter, i.e., quick re-entry into B_N after entering G_N. Under the regularity condition $\frac{\lambda_n}{\mu_n} \to \rho$, the asymptotic distribution of T_{VN} consists of two parts: the jitter term, and the exponential term.

§8.5. Limit Behavior of Other Exit Times for More General Chains.

The results of §8.4 may be extended to the ergodic exit time and quasi-stationary time for more general Markov chains. Consider the following setting.

Context 8.5A. Let $N(t)$ be an ergodic Markov chain in continuous time on a denumerably infinite state space $\mathcal{N}$. Let $\mathcal{N} = G_K + B_K$ be a sequence of nested partitions of $\mathcal{N}$, i.e., let $G_K \subset G_{K+1}$, all K, and let $E[T_{0B_K}] \to \infty$, as $K \to \infty$ as in §8.2. Let $P(G_K)$, the ergodic probability for set G_K, go to one as $K \to \infty$.

In this setting a variety of simple results may be demonstrated.

Theorem 8.5B. Consider Context 8.5A. Let m be any fixed state in G_K. Then $E[T_{mB_K}] \to \infty$, as $K \to \infty$. Moreover,

$$\lim_{K\to\infty} \frac{E[T_{mB_K}]}{E[T_{0B_K}]} \to 1. \tag{8.5.1}$$

The proof follows from the triangle inequality of §7.7 and an auxiliary argument using the idea of aggregating states. For the partition $\mathcal{N} = G_K + B_K$, let the chain $N(t)$ be modified by aggregating all the states of B_K into a single state n_K^*, so that $\mathcal{N}^* = G_K + n_K^*$ and let samples reaching n_K^* be replaced immediately at state 0. Since $e_0 > 0$, one has $0 \in G_K$, all $K > K_0$. Let the modified process be denoted by $N_{1K}^*(t)$. Then from the triangle inequality,

$$\overline{T}_{mn_K^*}^{(K)} \le \overline{T}_{m0}^{(K)} + \overline{T}_{0n_K^*}^{(K)}, \tag{8.5.2}$$

where $\overline{T}_{m0}^{(K)}$ is the mean time from state m to state 0 for $N_{1K}^*(t)$. We note that $\overline{T}_{mn_K^*}^{(K)} = \overline{T}_{mB_K}$ and $\overline{T}_{0n_K^*}^{(K)} = \overline{T}_{0B_K}$. Moreover, $\overline{T}_{m0}^{(K)} \le \overline{T}_{m0}$ since samples reaching 0 from m through B_K have their transit times reduced. It follows that the triangle inequality is valid for the set B_K, i.e.,

$$\overline{T}_{mB_K} \le \overline{T}_{m0} + \overline{T}_{0B_K}. \tag{8.5.3}$$

Clearly the symmetry of this argument in m and 0 permits these two states to be interchanged in (8.5.3), i.e.,

$$\overline{T}_{0B_K} \le \overline{T}_{0m} + \overline{T}_{mB_K}. \tag{8.5.4}$$

If we divide (8.5.3) and (8.5.4) by $\overline{T}_{0B_K}$ we have

$$1 - \frac{\overline{T}_{0m}}{\overline{T}_{0B_K}} \le \frac{\overline{T}_{mB_K}}{\overline{T}_{0B_K}} \le 1 + \frac{\overline{T}_{m0}}{\overline{T}_{0B_K}} \tag{8.5.5}$$

and the theorem is obtained by letting $K \to \infty$. □

Theorem 8.5C. In the context of 8.5A, let T_{EK} be the ergodic exit time from set G_K, with

$$\overline{T}_{EK} = \frac{\sum_{G_K} e_m \overline{T}_{mB_K}}{\sum_{G_K} e_m} . \tag{8.5.6}$$

Then

$$\frac{\overline{T}_{EK}}{\overline{T}_{0B_K}} \to 1, \text{ as } K \to \infty. \tag{8.5.7}$$

The proof again follows from the dominated convergence theorem, as for Theorem 8.3C. Clearly the state zero may be replaced by any state n in $\mathcal{N}$ for $K > n$, $K \to \infty$.

Corollary 8.5D. In the context of 8.5A, if $N(t)$ is reversible in time and T_{QK} is the quasi-stationary exit time from G_K, then

$$\overline{T}_{QK}/\overline{T}_{qB_K} \to 1, \quad K > q, \quad K \to \infty, \tag{8.5.8}$$

for every state q in $\mathcal{N}$.

The proof is as in §8.3 and is unchanged.

The discussion of the limit behavior in distribution of the different exit times is slightly different in proof. Again the ideas of §7.7 provide the key.

In the context of 8.5A we again consider the sequence of ergodic exit times T_{EK}. A lemma is needed.

Lemma 8.5F. Let m be any fixed state in $\mathcal{N}$ and let $\sigma_{mB_K}(s) = E[e^{-sT_{mB_K}}]$. Then

$$\frac{T_{mB_K}}{\overline{T}_{0B_K}} \overset{d}{\to} Y, \tag{8.5.9}$$

where $F_Y(x) = 1 - e^{-x}$.

Proof of Lemma: From (7.7.4a,b) we have for any state 0, α, and β of an ergodic chain

$$\sigma_{0\alpha}(s) \geq \sigma_{0\beta}(s)\sigma_{\beta\alpha}(s), \quad s > 0. \tag{8.5.10}$$

Consider the modified process $N^*_{1K}(t)$ in the proof of Theorem 8.5B. Then for $\sigma^{(K)}_{0n^*_K}(s) = E[e^{-sT_{0n^*_K}}]$ we have

$$\sigma^{(K)}_{0n_{K*}}(s) \geq \sigma^{(K)}_{0m}(s)\ \sigma^{(K)}_{mn^*_K}(s).$$

Since $T^{(K)}_{0m} \prec T_{0m}$, one has $\sigma^{(K)}_{0m}(s) \geq \sigma_{0m}(s)$, $s > 0$. Hence from (8.5.10) as for Theorem 8.5B

$$\sigma_{0B_K}(s) \geq \sigma_{0m}(s)\sigma_{mB_K}(s), \quad s > 0 \tag{8.5.11}$$

and

$$\sigma_{0B_K}\left(\frac{s}{\overline{T}_{0B_K}}\right) \geq \sigma_{0m}\left(\frac{s}{\overline{T}_{0B_K}}\right)\ \sigma_{mB_K}\left(\frac{s}{\overline{T}_{0B_K}}\right). \tag{8.5.12}$$

Again from the symmetry of the argument in m,0

$$\sigma_{mB_K}\left(\frac{s}{\overline{T}_{0B_K}}\right) \geq \sigma_{m0}\left(\frac{s}{\overline{T}_{0B_K}}\right)\ \sigma_{0B_K}\left(\frac{s}{\overline{T}_{0B_K}}\right). \tag{8.5.13}$$

When $K \to \infty$, $s/\overline{T}_{0B_K} \to 0$. Hence from (8.5.12), (8.5.13) and

$$\sigma_{0B_K}\left(\frac{s}{\overline{T}_{0B_K}}\right) \to \frac{1}{1 + s}, \quad K \to \infty$$

from Theorem 8.2B, the lemma follows.

Theorem 8.5G. In Context 8.5A, let T_{EK} be the ergodic exit time from G_K. Then

$$\frac{T_{EK}}{\bar{T}_{EK}} \xrightarrow{d} Y, \quad \text{as} \quad K \to \infty \tag{8.5.14}$$

where $F_Y(x) = 1 - e^{-x}$.

Proof: The proof follows from $\bar{T}_{EK} \sim \bar{T}_{0B_K}$ (Theorem 8.5C),

$$\sigma_{EK}\left(\frac{s}{\bar{T}_{0B_K}}\right) = \frac{\sum_{G_K} e_m \, \sigma_{mB_K}\left(\frac{s}{\bar{T}_{0B_K}}\right)}{\sum_{G_K} e_m}, \tag{8.5.15}$$

the dominated convergence theorem, and Lemma 8.5F exactly as in the proof of Theorem 8.4A.

Corollary 8.5H. In the context of 8.5A, let T_{QK} be the quasi-stationary exit time from G_K. Suppose further that $N(t)$ is reversible in time and that there is a perfect ("brand new" in reliability theory) state 0 such that

$$T_{0B_K} \succ T_{mB_K} \text{ for all } m \in \mathcal{N}.$$

Then

$$T_{QK}/\bar{T}_{QK} \xrightarrow{d} Y; \quad F_Y(x) = 1 - e^{-x}. \tag{8.5.16}$$

The proof is identical to that of Corollary 8.4B, and follows from $T_0 \succ T_Q \succ T_E$. The property $T_{0B_K} \succ T_{mB_K}$ is a "new better than used" property which has been demonstrated, for example, for reliability systems with independent Markov components, by S. Ross.[†]

[†] Ross, S. "On Time to First Failure in Multicomponent Exponential Reliability Systems," ORC Report 74-8, Operations Research Center, Univ. of Calif., Berk. (1974).

§8.6. Strongly Stable Chains; Jitter; Estimation of the Failure Time Needed for The Exponential Approximation.

In the study of birth-death processes in §8.3 we saw that when $\lambda_n/\mu_n \to \rho$ and when the mean transit time $\overline{T}_{0K} \to \infty$ as $K \to \infty$, the asymptotic sojourn time T_{VK} on $G_K = \{0,1,2, \ldots, K\}$ has the limiting behavior

$$\lim_{K\to\infty} \frac{\overline{T}_{0K}}{\overline{T}_{VK}} = \lim_{K\to\infty} \frac{\overline{T}_{EK}}{\overline{T}_{VK}} = (1 - \rho)^{-1}.$$

When ρ goes to zero, the ergodic exit time and ergodic sojourn times are asymptotically equal. Formally we may define a "jitter factor" J_K for *any* ergodic process and any partition $\mathcal{N} = G_K + B_K$ by

$$J_K = \frac{\overline{T}_{EK}}{\overline{T}_{VK}} . \tag{8.6.1}$$

As we have seen, for every ergodic time-reversible chain, $J_K \geq 1$, (6.8.8); and for every ergodic chain $J_K \geq 1/2$, (6.8.4). The jitter when $P(G_K) \approx 1$ corresponds to clustering of the epochs at which the process crosses from G_K to B_K, i.e., to multiple crossings in an interval small compared to the mean ergodic exit time.

The jitter factor J_K may be viewed in terms of the ratio between two flow rates. If $N(t)$ is the process of interest, let $N_{MK}(t)$ be the modified process obtained from $N(t)$ by replacing samples reaching B_K in G_K at state m with probability $e_m/P(G_K)$. Then the ratio

$$i_{G_K B_K} / i^M_{G_K B_K}$$

between the ergodic flow rates for the unmodified and modified processes is from (6.7.11) and a simple renewal argument,

$$\frac{i_{G_K B_K}}{i^M_{G_K B_K}} = \frac{P(G_K)/\bar{T}_{VK}}{1/\bar{T}_{EK}} = J_K P(G_K). \tag{8.6.2}$$

For certain processes, such as the birth-death process above with $\rho = 0$, one has in the context of 8.5A,

$$\lim_{K\to\infty} J_K = 1$$

Such processes are characterized by a growing bias back towards the normal domain of the state space as K increases. A chain having this behavior may be said to be "strongly stable." The mean ergodic sojourn time $\bar{T}_V$ may be used to approximate the mean ergodic exit time T_E or the mean time $\bar{T}_{0B}$ from a state 0 deep in G only to the extent that the process behaves in a strongly stable manner.

<u>Example 8.6A</u>. Suppose a system consists of a large number K of components, each governed by exponential failure times and exponential repair times. The components differ, however, in their failure and repair rates. It is known that the failure rates $\lambda_j < \lambda_{Max}$, and the repair rates $\mu_j > \mu_{Min}$, and that $\lambda_{Max}/\mu_{Min} < 1/2$. By comparison with the process for which $\lambda_j = \lambda_{Max}$, $\mu_j = \mu_{Min}$ it is easily seen that P{another failure before another repair | k components are working} $< k\lambda_{Max}\{(K-k)\mu_{Min} + k\lambda_{Max}\}^{-1} < k/(2K - k)$. As k decreases the process is biased more strongly to larger k. Hence the process has a strongly stable character, and the failure time may be estimated by the expected sojourn time.

In a particular reliability problem, modeled by an ergodic chain it may be clear that jitter is present. But quantification of the jitter factor may be difficult. Direct calculation of the mean time to failure from the perfect state may not be feasible. The estimate $\overline{T}_0 \approx \overline{T}_E$ is implicit in the limit theorem of §8.2 when high reliability is present. If, further, it is known that $\overline{T}_E \approx \overline{T}_Q$ from considerations of time-reversibility, say, an estimate of $\overline{T}_Q$ may be appealing.

One way of estimating $\overline{T}_Q$ is from perturbation theory. Suppose the process $N(t)$ is modelled by a time-reversible ergodic chain. As shown in §6.6, the quasi-stationary exit time has a survival function $e^{-\nu\tau[1-r]}$ where r is the maximal eigenvalue of the substochastic matrix a_ν with elements $a_{\nu nn} = 1 - \nu_n/\nu$; $a_{\nu mn} = \nu_{mn}/\nu$ (cf. §2.1). The time-reversible chain assures that $e_D a_\nu$, and hence $e_D^{1/2} a_\nu e_D^{-1/2}$ is real symmetric. Its maximal eigenvalue may then be estimated with the help of the kind of perturbation theory employed in quantum mechanics for the calculation of the principal eigenvalue of a self-adjoint operator. Such techniques have been highly developed in physics and may be appropriate to the study of the failure times of reliable ergodic systems.

§8.7. A Measure of Exponentiality in the Completely Monotone Class of Densities.

To be able to say how good the exponential approximation is, a simple measure of exponentiality is needed. When time-reversibility is present for the chain of interest, complete monotonicity is often at hand in the descriptive distributions of interest. In this class, a good simple measure is available [39].

As we have seen in §5.6, Bernstein's representation (Theorem 5.4B) may be regarded as stating that a completely monotone density is a scale mixture of exponential densities, i.e., the random variable has the form

$$X = YW \tag{8.7.1}$$

as far as its distribution is concerned. Here Y has density e^{-x} and W is an arbitrary positive mixing random variable. Consequently, (Theorem 5.6A) $(\sigma^2/\mu^2)_X \geq 1$. We will now see that the measure

$$\theta_X = \left(\frac{\sigma^2}{\mu^2}\right)_X - 1 \tag{8.7.2}$$

is a distance to pure exponentiality in a metric space sense, for those completely monotone densities that have finite second moments. It will be useful and convenient to enlarge the set of distributions of interest to permit finite support at 0. We are also interested only in a measure of the shape of the distribution, and restrict our attention correspondingly to those distributions $F_X(x)$ for which $EX = EW = 1$. Hence, let $\mathscr{D}$ be the space

$$\mathscr{D} = \{F_X\colon X = YW;\ Y,W \text{ independent};\ W \geq 0;\ EX = EW = 1\}. \tag{8.7.3}$$

The perfect exponential distribution in this class has $W = 1$. We define on this space the distance function

$$\rho(F_A, F_B) = \int (\omega - 1)^2 |\mu_{W_A}(d\omega) - \mu_{W_B}(d\omega)|, \tag{8.7.4}$$

where F_A is associated with W_A and its probability measure $\mu_{W_A}(d\omega)$.

$$\text{a)}\quad \rho(F_A,F_B) \geq 0; \quad \rho(F_A,F_B) = 0 \Leftrightarrow \mu_{W_A} = \mu_{W_B}$$

$$\text{b)}\quad \rho(F_A,F_B) = \rho(F_B,F_A) \tag{8.7.5}$$

$$\text{c)}\quad \rho(F_A,F_C) \leq \rho(F_A,F_B) + \rho(F_B,F_C),$$

i.e., that ρ satisfies all the requirements of a distance function. We note from (8.7.1) that when W_Y has all support at $\omega = 1$,

$$\rho(F_A, F_Y) = \int (\omega-1)^2 \mu_{W_A}(d\omega) = \left(\frac{\sigma^2}{\mu^2}\right)_{W_A}, \tag{8.7.6}$$

i.e., one has

$$2\rho(F_X,F_Y) = \left(\frac{\sigma^2}{\mu^2}\right)_X - 1. \tag{8.7.7}$$

It is clear from (8.7.6) why $\left(\frac{\sigma^2}{\mu^2}\right)_X - 1$ is a measure of exponentiality. It is the squared coefficient of variation of W_X, i.e., is a measure of the purity of the spectral measure $F_{W_X}(\omega)$ appropriate to Bernstein's representation. When the limit theorem of §8.5 says that $T_{EK}/\overline{T}_{EK} \to Y$, and the process $N(t)$ is time-reversible, the complete monotonicity of $s_{EK}(\tau)$ assures that the spectral measures F_{W_K} for $T_{EK}/\overline{T}_{EK}$ are converging to the measure with all support at $\omega = 1$.

We note that when $W \geq 0$, and $EW^2 < \infty$. the r.v. W has $\left(\frac{\sigma^2}{\mu^2}\right)_W = 0$ iff W has all support at 1. Correspondingly, a sequence f_{K_n} of completely monotone densities rescaled to expectation one converges to exponentiality iff $(\sigma^2/\mu^2)_{X_n} - 1 \to 0$. This provides a tool for the demonstration of convergence in distribution to exponentiality.

§8.8. An Error Bound for Departure from Exponentiality in the Completely Monotone Class.

The measure $\theta_X = (\sigma^2/\mu^2)_X - 1$ of §8.7 for the exponentiality of a density f_X in the completely monotone class CM permits one to say meaningfully that a density f_1 in the class CM is "more exponential than" a second such density f_2. Clearly it would be desirable to express the departure of f_X from exponentiality when its measure is θ_X in a more concrete way, i.e., to give an error bound. The availability of such a bound has been demonstrated by C. C. Heyde[†] for any mixture class $\{f_X: X = YW\}$ when Y and W are independent and have finite first moments. The special case of interest to us is described by

Theorem 8.8A. Let $X = YW$, where Y has c.d.f. $1 - e^{-x}$, and W is a non-negative r.v. independent of X with $EW = 1$. Then

$$|P[X > x] - e^{-x}| \le 8\pi^{-1}\sqrt{3}\, 2^{-1/4}(\sigma_X^2 - 1)^{1/4} \qquad (8.8.1)$$

for all positive X.

Proof: The proof follows from Lemma 2 of Feller II, [15], (1966), on p. 512 and

$$\begin{aligned} |Ee^{itYW} - Ee^{itY}| &\le E|\{e^{itY(W-1)} - 1\}| \\ &= E\left\{\left|\frac{e^{itY(W-1)} - 1}{itY(W-1)}\right| |t|\,|Y(W-1)|\right\} \\ &\le |t|\; E|W - 1|. \end{aligned}$$

Hence from the lemma in Feller,

[†]See Heyde [17], and related papers by Heyde and Leslie [18] and P. Hall [16].

$$|P[X > x] - e^{-x}| \leq \lim_{T>0} \left\{ \frac{2T}{\pi} E|W - 1| + \frac{24}{\pi T} \right\}$$

$$= 8\pi^{-1}\sqrt{3}\, E^{1/2}|W - 1| \leq 8\pi^{-1}\sqrt{3}\, \{E(W - 1)^2\}^{1/4}.$$

But $EW = 1 \Rightarrow E(W - 1)^2 = \sigma_W^2 = \dfrac{\sigma_X^2 - 1}{2}$. This gives (8.8.1).

It is quite likely that the constant $8\pi^{-1}\sqrt{3}\, 2^{-1/4}$ can be improved.

§8.9. The Exponential Approximation for Time-Reversible Systems.

For a given multi-component real world system modelled by a time-reversible ergodic Markov chain $N(t)$ the crucial problem will be the justification of the exponential approximation in the manner described at the end of §8.1. Some tools for this justification may be of interest.

Consider the passage time T_{AB} from some state A in the good set G to the bad set B with p.d.f. $s_{AB}(\tau)$. From (8.1.2) and the subsequent discussion we see that T_{AB} is the sum of two independent times, i.e.,

$$T_{AB} = T_C + T_{AB}^{+}, \tag{8.9.1}$$

where T_C is the multicycle time ending at the last return to state A before the direct passage to set B and T_{AB}^{+} is the time from that return to set B. In this notation (8.1) may be rewritten as $\sigma_{AB}(s) = \sigma_C(s)\sigma_{AB}^{+}(s)$.

We first examine the structure of the density of T_C, $s_C(t)$. From (8.1.2) and (8.1.3) its transform is

$$\sigma_C(s) = \frac{1 - \theta_A^{+}}{1 - \theta_A^{+} \dfrac{\nu_A}{\nu_A + s} \sigma_V^{+}(s)}, \tag{8.9.2}$$

where $s_\nu^+(\tau)$ is the sojourn time density in the exterior of {A}. Let $N^*(t)$ be the transient chain obtained from $N(t)$ by making the bad set B absorbing, and let

$$p_{AA}^*(t) = P[N(t) = A;\ N(t') \in G,\ 0 \le t' \le t \mid N(0) = A].$$

Then via the customary continuity argument one has

$$\frac{d}{dt} p_{AA}^*(t) = -\nu_A p_{AA}^*(t) + \nu_A p_{AA}^*(t) * \theta_A^+ s_V^+(t), \qquad (8.9.3)$$

where θ_A^+ is the probability of return to A as in §8.1. From §3.5 we note that $p_{AA}^*(t)$ is completely monotone. It is also quickly seen that $s_V^+(t)$ is completely monotone. For, from (8.9.3) one has

$$\theta_A^+ \sigma_V^+(s) = \frac{(s + \nu_A)\pi_{AA}^*(s) - 1}{\nu_A \pi_{AA}^*(s)} . \qquad (8.9.4)$$

The modified process $N^*(t)$ is time-reversible (§3.5), and $\pi_{AA}^+(s)$ has zeros only on the negative real axis (cf. §5.1). At these zeros $(s + \nu_A)\pi_{AA}^*(s) - 1 = 0 - 1 = -1$. The residues at these zeros all have the same sign (cf. §5.1) and the complete monotonicity of $s_V^+(\tau)$ follows.

Let $s_0(\tau) = p_{AA}^*(\tau)/\int p_{AA}^*(\tau)dt$, so that $s_0(\tau)$ is a p.d.f. Then one has the decomposition, from (8.9.2) and (8.9.4),

$$s_C(\tau) = (1 - \theta_A^+)\delta(\tau) + \theta_A^+ s_V^+(\tau) * s_0(\tau), \qquad (8.9.5)$$

where $\delta(\tau)$ is the generalized density with unit support at $\tau = 0$. When $\theta_A^+ \approx 1$, almost all of the support of $s_C(\tau)$ is in the term on the right, and the dominant term there is $s_0(\tau)$. This may be inferred from (8.9.4) which gives

$$\overline{T}_0 = -\frac{d}{ds}\left[\frac{\pi^*_{AA}(s)}{\pi^*_{AA}(0)}\right]_{s=0} = \frac{\frac{1}{\nu_A} + \theta^+_A \overline{T^+_V}}{1 - \theta^+_A} \tag{8.9.6}$$

so that when $\theta^+_A \approx 1$, $T_0 >> \overline{T^+_V}$. If further, $\overline{T^+_{AB}} << T_C$; then $\overline{T}_0$ is the major term in T_{AB}, as seen from (8.9.1) and (8.9.5). The complete monotonicity of $s_0(\tau)$ is vital, for this then permits the use of $\left[\left(\frac{\sigma^2}{\mu^2}\right) - 1\right]_{T_0}$ of §8.7 and §8.8 as a measure of the exponentiality of $s_\theta(\tau)$ and hence of $s_{AB}(\tau)$.

The reader will verify from (8.9.2) and (8.9.5) that

$$\sigma_0(s) = \frac{\nu_A}{\nu_A + s}\,\sigma_C(s), \tag{8.9.7}$$

i.e., that T_0 is the time from state A to the last departure from A before hitting B.

For further evaluation of $[(\sigma^2/\mu^2) - 1]_{T_0}$ we proceed as follows. Let $\phi_X(s) = Ee^{-sX}$. Then, from $\left[\frac{d^2}{ds^2}\log\phi_X(s)\right]_{s=0} = \sigma^2_X$; $\left[-\frac{d}{ds}\log\phi_X(s)\right]_{s=0} = \mu_X$ we have for $\phi_X = \pi^+_{AA}(s)/\pi^+_{AA}(0)$ and $\pi^+_{AA}(s) = [s + \nu_A - \nu_A\theta^+_A\sigma^+_V(s)]^{-1}$,

$$\sigma^2_0 = \frac{(1 + \nu_A\theta^+_A\overline{T^+_V})^2 + \nu^2_A\theta^+_A(1 - \theta^+_A)\,(\overline{T_V^{+2}})}{\nu^2_A(1 - \theta^+_A)^2} \tag{8.9.8}$$

and

$$\mu^2_0 = \frac{(1 + \nu_A\theta^+_A\overline{T^+_V})^2}{\nu^2_A(1 - \theta^+_A)^2}\,. \tag{8.9.9}$$

It follows that

$$\frac{\sigma^2_0}{\mu^2_0} - 1 = \frac{\nu^2_A\theta^+_A(1 - \theta^+_A)(\overline{T_V^{+2}})^2}{[1 + \nu_A\theta^+_A\overline{T^+_V}]^2}.$$

If the set B is rarely visited and $\theta^+_A \approx 1$ we may replace $(\overline{T_V^{+2}})$ by $\overline{T^2_V}$, and $\overline{T^+_V}$ by $\overline{T_V}$, where T_V is the sojourn time of $N(t)$, the parent process, on $\mathcal{N} - \{A\}$. We then have the estimate, using (7.6.12)

$$\frac{\sigma_0^2}{\mu_0^2} - 1 \approx \frac{\nu_A^2(1 - \theta_A^+)\,\overline{T_V^2}}{[1 + \nu_A \overline{T}_V]^2} = (1 - \theta_A^+)\nu_A^2 e_A^2\, \overline{T_V^2}. \tag{8.9.10}$$

From (7.6.12) and (7.6.13) we see that when $\overline{T}_{BA} << \overline{T}_{AB}$ (see §8.10)

$$\frac{\sigma_0^2}{\mu_0^2} - 1 \approx \frac{\nu_A \overline{T_V^2}}{\overline{T}_{AB}} = \frac{2(1 - e_A)}{e_A} \frac{\overline{T}_E(\bar{A})}{\overline{T}_{AB}} . \tag{8.9.11}$$

The quantity $\overline{T}_E(\bar{A})/\overline{T}_{AB}$ will be very small in a highly reliable system.

The explicit quantification of $(\sigma_0^2/\mu_0^2) - 1$ for a system with independent Markov components is possible from (8.9.10).

The quantity $\overline{T_V^2}$ may be expressed in terms of $p_{AA}(t)$ from $\pi_{AA}(s) = [s + \nu_A - \nu_A\sigma_V(s)]^{-1}$. This may be rewritten in the form $\alpha(s)[1 + \beta(s)] = \beta(s)$, where $\alpha(s) = 1 - s\pi_{AA}(s)$ and $\beta(s) = \nu_A s^{-1}[1 - \sigma_V(s)]$. From the behavior at $s = 0$, we have

$$\nu_A \frac{\overline{T_V^2}}{2} = \frac{\int (p_{AA}(t) - e_A)\, dt}{e_A} \nu_A \overline{T}_V \tag{8.9.12}$$

and $\nu_A \overline{T}_V = (1 - e_A)e_A^{-1}$. Again $p_{AA}(t)$ is available from the independence of the components. The key entry $(1 - \theta_A^+)$ may be found via (7.6.13) and the methods of §6.9, §8.5 and §8.6.

The verification that $\overline{T}_{AB}^+$ is small for a reliable system might proceed as follows. Let the states of the set B be aggregated as in §3.5 to a single state designated B. As we have seen, the passage time to the set B and that to state B for the modified process are identical. For the modified process, from (7.6.6) and (7.6.7) we have for $\sigma_C(s) = E[\exp(-sT_C)]$ and $\sigma^+_{AB}(s) = E[\exp(-sT^+_{AB})]$, using (7.6.13)

$$\sigma^+_{AB}(s) = \frac{\sigma_{AB}(s)\left[\dfrac{1-\rho_A(s)}{s\overline{T}_A}\right]}{\dfrac{1-\sigma_{AB}(s)\sigma_{BA}(s)}{s(\overline{T}_{AB}+\overline{T}_{BA})}} = \frac{\alpha(s)\beta(s)}{\gamma(s)} \qquad (8.9.13a)$$

$$\sigma_C(s) = \frac{\dfrac{1-\sigma_{AB}(s)\sigma_{BA}(s)}{s(\overline{T}_{AB}+\overline{T}_{BA})}}{\dfrac{1-\rho_A(s)}{s\overline{T}_A}} = \frac{\gamma(s)}{\beta(s)}\ , \qquad (8.9.13b)$$

where $\alpha(s)$, $\beta(s)$, and $\gamma(s)$ are all characteristic functions. From $-[\log \sigma_X]^1_0 = \mu_X$ we have $\overline{T}^+_{AB} = \mu_\alpha + \mu_\beta - \mu_\gamma$ and $\overline{T}_C = \mu_\gamma - \mu_\beta$. The negligibility of $\overline{T}^+_{AB}$ with respect to $\overline{T}_C$ may be seen from

$$\frac{\overline{T}^+_{AB}}{\overline{T}_C} = \frac{\mu_\alpha - \mu_\gamma + \mu_\beta}{\mu_\gamma - \mu_\beta} = \frac{\dfrac{\mu_\beta}{\mu_\gamma} - \dfrac{\mu_\gamma - \mu_\alpha}{\mu_\gamma}}{1 - \dfrac{\mu_\beta}{\mu_\gamma}} \qquad (8.9.14)$$

and a) $\left|\dfrac{\mu_\beta}{\mu_\gamma}\right| \ll 1$; b) $\left|\dfrac{\mu_\gamma - \mu_\alpha}{\mu_\alpha}\right| \ll 1$. The quantity μ_β is the mean time until the next return to state A at stationarity. The quantity μ_γ is the mean time until the next return to state A *through state* B at stationarity. Hence $\mu_\gamma \approx \mu_\alpha = \overline{T}_{AB}$, μ_β/μ_γ is small, and $|\mu_\gamma - \mu_\alpha|/\mu_\gamma \approx \overline{T}_{BA}/\overline{T}_{AB}$. We will soon see (§8.10) that $\overline{T}_{BA}/\overline{T}_{AB}$ is small so that both a) and b) are valid. To quantify the negligibility explicitly for a particular system, one proceeds as earlier. The quantity μ_β is obtained in a manner similar to that for $\overline{T^2_V}$ given above.

§8.10. A Relaxation Time of Interest (Joint with D. R. Smith).

A key theme in the above has been the use of the inequality $\bar{T}_{AB} >> T_{BA}$. The validity of this inequality for systems with independent Markov components when visits to B are rare stems from the independence. If component j has hazard rates μ_j and λ_j, it has a relaxation time or forgetting time given by $(\mu_j + \lambda_j)^{-1}$. For a system with K components the forgetting time is essentially equal to $\text{Max}\{(\lambda_j + \mu_j)^{-1}\}$. Visits to the bad set correspond to coincidences, i.e., to superpositions of failures which dissolve in this forgetting time. The time to the set of the bad coincidences, on the other hand, is very large when the individual failures are rare.

To place these ideas on more solid ground, the following relaxation time is introduced. Let $\underline{N}(t)$ be the vector process with components 0 and 1 indexing the set of working components. Let $\phi(\underline{n})$ be the structure function of the system, i.e., $\phi(\underline{n}) = 1$, $\underline{n} \in G$; $\phi(\underline{n}) = 0$, $\underline{n} \in B$. Then $\Phi(t) = \phi(\underline{N}(t))$ is the indicator process for the good set, and

$$r_\phi(\tau) = \text{cov}[\Phi(t), \Phi(t + \tau)], \tag{8.10.1}$$

its covariance function, is completely monotone, as shown in Theorem 7.4A. The relaxation time for the system is then defined by

$$T_{REL} = \int_0^\infty \frac{r_\phi(\tau)\, d\tau}{r_\phi(0)} . \tag{8.10.2}$$

The function $r_\phi(\tau)/r_\phi(0)$ has the appearance of a survival function. In a sense it is the survival function for the dependence of the system on its past. For a single component two-state system with failure rate μ and repair rate λ, one finds easily $T_{REL} = (\lambda + \mu)^{-1}$ as the reader will verify.

As for (7.4.5) we may write easily

$$\frac{\int_0^\infty r_\phi(\tau)\, d\tau}{r_\phi(0)} = \frac{\underline{e}_G^T Z \underline{1}_G}{\underline{e}_G^T(I - \underline{1}\underline{e}^T)\underline{1}_G} = \frac{\underline{1}_G^T(e_D Z)\underline{1}_G}{\underline{1}_G^T(e_D - \underline{e}\underline{e}^T)\underline{1}_G}, \quad (8.10.3)$$

where $\underline{1}_G$ is the indicator vector for the states of the good set G and Z is the fundamental matrix of Chapter 7. The spectral representation of Z given in (7.2.1) together with $I = \sum_1^K J_j$, permits one to rewrite (8.10.3) in the form (if there are K components)

$$T_{REL} = \frac{\sum_2^K \frac{1}{\alpha_j} \underline{1}_G^T(e_D J_j)\underline{1}_G}{\sum_2^N \underline{1}_G^T(e_D J_j)\underline{1}_G}, \quad (8.10.4)$$

where $\alpha_j > 0$ and $e_D J_j$ are symmetric dyads, and hence non-negative-definite. It follows at once that

$$T_{REL} \leq \max_{j \geq 2} \left(\frac{1}{\alpha_j}\right). \quad (8.10.5)$$

Clearly, for processes with independent components, the eigenvalues of $p(t)$ for $N(t)$ and hence the reciprocals of those for Z are given by

$$\alpha_j = \sum_{k=1}^{2^K} \delta_j(k)(\lambda_k + \mu_k), \quad \delta_j(k) = 0,1. \quad (8.10.6)$$

Hence, after the exclusion of the eigenvalue at 0, we have

$$T_{REL} \leq \max_k \left(\frac{1}{\lambda_k + \mu_k}\right).$$

This is valid for any structure function. The results are

quoted formally as a theorem.

Theorem 8.10A. Let $\underline{N}(t)$ be any ergodic finite vector process with independent Markov components $N_i(t)$. Let $\mathcal{N} = G + B$ be any parition of the state space, and let $\Phi(t)$ be the corresponding indicator process for set G. Then for the relaxation time $T_{REL}(G)$ defined above, and any partition,

$$T_{REL}(G) \leq \text{Max}\, T_{REL}(C_i),$$

where $T_{REL}(C_i)$ is the relaxation time of the ith component.

In the same manner we find

Theorem 8.10B. Let $N(t)$ be a finite, ergodic time-reversible process. Let

$$T_{REL} \overset{d}{=} \underset{\phi}{\text{Max}} \int_0^\infty \frac{r_\phi(\tau)}{r_\phi(0)}\, d\tau,$$

where ϕ is the indicator function for a subset of the state space. Then

$$T_{REL} = \underset{j\geq 2}{\text{Max}} \left[\frac{1}{\alpha_j}\right] = \frac{1}{\alpha^*},$$

where the α_j are as in (7.2.1). Moreover, for any function $f(n)$, we have for $\tau \geq 0$

$$\frac{\text{Cov}\,\{f[N(t)],\, f[N(t+\tau)]\}}{\text{Var}\{f[N(0)]\}} \leq e^{-\tau/T_{REL}}.$$

There follows at once from §7.5,

Corollary 8.10C. Let $S(t) = \int_0^t f[N(t')]dt'$. Then

$$\lim_{t\to\infty} \frac{\text{Var}\, S(t)}{t} \leq 2\left\{\sum_n e_n f^2(n)\right\} T_{REL}.$$

Chapter 9

Stochastic Monotonicity

§9.0. Introduction.

A Markov process $X(t)$ is said to be monotone if $P[X(t) > x \mid X(0) = y]$ increases with y for every fixed x. The monotonicity of operators governing processes was introduced by Kalmykov [20], Veinott [64], and Daley [10], and discussed further by O'Brien [56], Kirstein [49], Keilson and Kester [36], [37], and Whitt and Sonderman [65]. The property is simple and widespread, and lends itself to a variety of structural insights. In particular, it is basic to many interesting inequalities in reliability theory.

9.1. Monotone Markov Matrices and Monotone Chains [36].

Stochastic monotonicity takes it simplest form in discrete time chains. It has its origin in certain properties of finite matrices. Some definitions are needed.

Def. 9.1A. Let P_N be the set of stochastic vectors in N-space, i.e., let $P_N = \{\underline{p} : p_m \geq 0, \sum_1^N p_m = 1\}$, and let $\underline{p}$, $\underline{q}$ be elements of P_N. $\underline{p}$ is said to be larger stochastically than $\underline{q}$ or to dominate $\underline{q}$ if $P_n = \sum_n^N p_m \geq Q_n = \sum_n^N q_m$,

$n = 1,2, \ldots, N$. One then writes $\underline{p} \succ \underline{q}$. ($N$ may be finite or infinite.) For a discussion of stochastic ordering on the continuum, the reader may be interested in the important paper of E. L. Lehmann [51].

<u>Def. 9.1B</u>. Let S_N be the set of stochastic matrices in N-space. Let $a \in S_N$ have the row vector representation $a = [\underline{a}_1^T, \underline{a}_2^T, \ldots, \underline{a}_N^T]$ where $\underline{a}_j^T$ are stochastic vectors in P_N. The stochastic matrix a is said to be stochastically monotone if $\underline{a}_1^T \prec \underline{a}_2^T \prec \ldots \prec \underline{a}_N^T$. One writes $a \in \mathcal{M}_N$. From 9.1A this means that one must have A_{mn} non-decreasing in m for every n.

Let t_N be the lower triangular matrix given by

$$t_N = \begin{vmatrix} 1 & 0 & 0 & . & . & 0 & 0 \\ 1 & 1 & 0 & . & . & 0 & 0 \\ 1 & 1 & 1 & . & . & 0 & 0 \\ . & . & . & . & . & . & . \\ . & . & . & . & . & . & . \\ 1 & 1 & 1 & . & . & 1 & 0 \\ 1 & 1 & 1 & . & . & 1 & 1 \end{vmatrix} \tag{9.1.1}$$

Clearly t_N is non-singular, since all its eigenvalues are equal to one. The reader will verify that when N is finite its inverse is given by

$$t_N^{-1} = \begin{vmatrix} 1 & 0 & 0 & 0 & . & . & 0 \\ -1 & 1 & 0 & 0 & . & . & 0 \\ 0 & -1 & 1 & 0 & . & . & 0 \\ . & . & . & . & . & . & . \\ . & . & . & . & . & . & . \\ 0 & 0 & 0 & 0 & . & 1 & 0 \\ 0 & 0 & 0 & . & . & -1 & 1 \end{vmatrix} \tag{9.1.2}$$

Then the relation between $\underline{p}$ and its tail sum vector $\underline{P}$ is

$(\underline{p}t_N)_n = P_n$, i.e., $\underline{p}t_N = \underline{P}$. Moreover, for any $\underline{A}$,

$$t_N^{-1}\underline{A} = [A_1, A_2 - A_1, A_3 - A_2, \ldots, A_N - A_{N-1}]. \quad (9.1.3)$$

Hence if $A_0 \geq 0$, and $A_n\uparrow$ (i.e., A_n is non-decreasing as n increases), then $(t_N^{-1}\underline{A})_n \geq 0$, i.e., $t_N^{-1}\underline{A} \geq \underline{0}$. The Definition 9.1B may then be given the alternate form

$$a \in \mathscr{M}_N \Longleftrightarrow a \in S_N \text{ and } t_N^{-1} a t_N \geq 0 \quad (9.1.4)$$

since $t_N^{-1} a t_N = t_N^{-1}[\underline{a}_1^T, \underline{a}_2^T, \ldots \underline{a}_N^T] t_N = t_N^{-1}[\underline{A}_1^T, \underline{A}_2^T, \ldots \underline{A}_N^T] = [\underline{A}_1^T, \underline{A}_2^T - \underline{A}_1^T, \ldots \underline{A}_N^T - \underline{A}_{N-1}^T]$.

The following closure properties follow immediately from (9.1.4).

<u>Proposition 9.1C</u>. $\mathscr{M}_N$ is closed under multiplication and mixing, i.e.,

$$a \in \mathscr{M}_N,\ b \in \mathscr{M}_N \Rightarrow \begin{cases} 1) & ab \in \mathscr{M}_N \\ 2) & pa + qb \in \mathscr{M}_N, \text{ for all } p,q \geq 0, \\ & p + q = 1. \end{cases}$$

For closure under multiplication, for example, one has $t_N^{-1} ab t_N = (t_N^{-1} a t_N)(t_N^{-1} b t_N) \geq 0$, since the product of non-negative matrices is non-negative. It follows from Property 9.1C that

$$a \in \mathscr{M}_N \Rightarrow a^K \in \mathscr{M}_N, \text{ all positive integers } K \quad (9.1.5)$$

$$a \in \mathscr{M}_N \Rightarrow e^{-\theta[I-a]} = \sum_0^\infty e^{-\theta}\frac{\theta^k}{k!} a^k \in \mathscr{M}_N \quad (9.1.6)$$

$$a \in \mathscr{M}_N \Rightarrow (1-\theta)[I-\theta a]^{-1} = \sum_0^\infty \theta^k a^k \in \mathscr{M}_N, \quad (9.1.7)$$

$$0 \leq \theta < 1$$

Stochastic monotonicity draws its importance from a pair of dual properties characterizing such matrices. Basically, monotone matrices are those stochastic matrices which preserve stochastic order under post-multiplication. Specifically, $a \in \mathcal{M}_N$ if $\underline{p}^T \succ \underline{q}^T \Rightarrow \underline{p}^T a \succ \underline{q}^T a$. The equivalent dual property of monotone matrices is that such matrices preserve component monotonicity pre-multiplication, i.e., if $A_n \uparrow$, and $a \in \mathcal{M}_N$, $(a\underline{A})_n \uparrow$. These characterizations will now be proven.

In what follows we assume for simplicity that N is finite.

<u>Theorem 9.1D</u>. The following three propositions are equivalent.

a) $a \in \mathcal{M}_N$

b) $\underline{p} \succ \underline{q} \Rightarrow \underline{p}a \succ \underline{q}a$

c) $f_n \uparrow \Rightarrow (a\underline{f})_n \uparrow$.

<u>Proof</u>: 1) $a \Rightarrow b$: Let $a \in \mathcal{M}_N$. $\underline{p} \succ \underline{q} \Rightarrow \underline{p}t \geq qt \Rightarrow \underline{p}t - \underline{q}t \geq \underline{0} \Rightarrow (\underline{p} - \underline{q})t \geq \underline{0} \Rightarrow \underbrace{(\underline{p} - \underline{q})t}_{\geq \underline{0}}\underbrace{(t^{-1}at)}_{\geq 0} \geq \underline{0} \Rightarrow (\underline{p} - \underline{q})at \geq \underline{0} \Rightarrow$

$\underline{p}at \geq \underline{q}at \Rightarrow \underline{p}a \succ \underline{q}a$,

2) $a \Rightarrow c$. $f_n \uparrow \Rightarrow (f_n + p)\uparrow, > 0$ for some $p > 0 \Rightarrow t^{-1}(\underline{f} + p\underline{1}) \geq \underline{0} \Rightarrow (t^{-1}at)t^{-1}(\underline{f} + p\underline{1}) \geq \underline{0} \Rightarrow t^{-1}a(\underline{f} + p\underline{1}) \geq \underline{0}$. $t^{-1}(a\underline{f} + p\underline{1}) \geq \underline{0} \Rightarrow (a\underline{f})_n \uparrow$. The proofs that $b \Rightarrow a$ and $c \Rightarrow a$ are similar.

The characterization (9.1.4) of monotone matrices has strong implications for the eigenvalues of the matrix. We note that a and $t_N^{-1} a t_N$ have common eigenvalues. We further note that $t_N^{-1} a t_N$ always has the block structure

$$t_N^{-1} a t_N = \left[\begin{array}{c|cccc} 1 & A_2 & A_3 & \cdots & A_N \\ \hline 0 & & & & \\ 0 & & M & & \\ \vdots & & & & \\ 0 & & & & \end{array}\right]$$

where M is non-negative. The maximal eigenvalue of M is then the maximal eigenvalue of a when the eigenvalue one is removed. It follows that when a is monotone and ergodic, it is guaranteed to have a second real positive eigenvalue λ^* larger than the others. In particular, $a^k = J_1 + (a-J_1)^k$ and $\underline{p}_k = \underline{p}_0 a^k$ approaches $\underline{e}$ the ergodic vector in a non-oscillatory manner. For further information, see [36].

The extension of the notion of monotonicity to a chain governed by a monotone matrix a is natural.

Def. 9.1E. A finite Markov chain N_k in discrete time is said to be monotone if its governing matrix a is monotone.

The implications of monotonicity for the behavior of the process will be discussed in the next section.

§9.2. Some Monotone Chains in Discrete Time.

Stochastic monotonicity is present in most of the simple processes studied in probability theory. Several examples will now be given.

a) Spatially Homogeneous Processes Modified by Boundaries. The Lindley process arising in the theory of waiting time is a discrete time prototype of a large class of Markov processes studied in the context of dams, inventory and queues [23]. The Lindley process is governed by

$$X_{k+1} = \text{Max}\,[0, X_k + \xi_{k+1}], \tag{9.2.1}$$

where $(\xi_j)_1^\infty$ is a sequence of i.i.d. random variables. For Markov chains these have integer values. Let $p_n = P[\xi_j = n]$. Then the homogeneous process (§4.5) on $\mathcal{N} = \{\ldots -2,-1,0,1,2\ldots\}$ is governed by $a_H = [a_{mn}]$ with $a_{mn} = p_{n-m}$. The tail probabilities for a_H are $A_{mn} = \sum_n^\infty a_{mr} = P_{n-m}$ where $P_n = \sum_n^\infty p_j$. Clearly P_{n-m} increases with m. Now consider the Lindley process which is the homogeneous process modified by a retaining boundary at zero (§4.5). Then, as the reader will easily see, a_L for the Lindley process has components $a^*_{mn} = a_{mn}$, when $m \geq 0$, $n \geq 1$, $a^*_{m0} = \sum_{-\infty}^{0} a_{mr}$. Consequently, for the tail sums one has $A^*_{mn} = A_{mn}$, $m \geq 0$, $n \geq 1$, $A^*_{m0} = 1$. Consequently, A^*_{mn} increases with m for every n, and a^*_L is monotone. For a second retaining boundary at $n = N$, one has the identical tail sum matrix. Only the state space is restricted. Clearly the governing matrix is again monotone.

If the boundary on the left is absorbing the first row becomes $1,0,0\ \ldots\ 0$. If the boundary on the right is absorbing the last row becomes $1,1,1,\ \ldots$. Other elements are unaffected. One concludes that modification by an absorbing or retaining boundary does not alter the monotonicity of the governing matrix a.

b) <u>The Galton-Watson Process with Immigration</u>. Let a population process be given, such that if at epoch k there are $N_k = n$ individuals present, the population at epoch $k+1$ is $N_{k+1} = \xi_{1k} + \xi_{2k} + \ldots + \xi_{nk} + \zeta_k$, where ξ_{ik}, ζ_k, $i = 1,2,\ \ldots,\ n$, are independent r.v.'s, with ξ_{ik} i.i.d. for fixed i. ξ_{ik} represents the number of individuals associated with the ith progenitor in the kth generation. ζ_k is the random number of immigrants. The p.g.f. of N_{k+1} given $N_k = n$ is therefore $g_n(u) = \sum_0^\infty a_{nm}u^m = \pi_\xi^n(u)\cdot\pi_\zeta(u)$, where $\pi_\xi(u)$, and $\pi_\zeta(u)$ are p.g.f.'s of ξ and ζ, respectively. Hence $g_{n+1}(u) = g_n(u)\cdot\pi_\xi(u)$. The random variable ξ is non-negative. The probability vector $\underline{a}_{n+1} = (a_{n+1,m})_0^\infty$ therefore dominates $\underline{a}_n$, and $\mathbf{a}$ is monotone.

Suppose that N_k is monotone. Its state probability vector $\underline{p}_k$ has $\underline{p}_k = \underline{p}_0\mathbf{a}^k$. Let $\underline{p}_0 = (1,0,0, \ldots, 0)$. Then $\underline{p}_0 \prec \underline{q}$ for *any* $\underline{q}$ in P_N. Hence $\underline{p}_0 \prec \underline{p}_0\mathbf{a}$ and from Theorem 9.1D, one then has $\underline{p}_0\mathbf{a} \prec \underline{p}_0\mathbf{a}^2$, etc. It follows that

$$\underline{p}_0 \prec \underline{p}_1 \prec \underline{p}_2 \prec \cdots, \tag{9.2.1}$$

i.e., the state probability vector increases in time.

We note that if $\underline{a}^T \prec \underline{b}^T$, and h_n increases with n, then $\underline{b}^T\,\underline{h} - \underline{a}^T\,\underline{h} = (\underline{b}^T - \underline{a}^T)\underline{h} = (b^T - a^T)t\,t^{-1}\underline{h} = (\underline{B}^T - \underline{A}^T)(t^{-1}\underline{h}) \geq 0$. ($\underline{B}^T - \underline{A}^T \geq 0$, and $t^{-1}\underline{h} \geq 0$.) Formally,

Prop. 9.2.1. If $\underline{a}^T \prec \underline{b}^T$, and $h_n \uparrow_n$, then $\Sigma h_n a_n \leq \Sigma h_n b_n$. Consequently,

Theorem 9.2.2. Let N_k be monotone on $\mathscr{N} = \{0,1,2, \ldots\}$. Then, if $h(n)\uparrow_n$, one has when $N_0 = 0$,

$$E[h(N_k) \mid N_0 = 0] \uparrow_k. \tag{9.2.2}$$

In particular, $E[N_k^K \mid N_0 = 0] \uparrow_k$ for all $K \geq 0$, and $P[N_k > m \mid N_0 = 0]\uparrow_k$ for all m.

If in the context of Theorem 9.2.2 the process N_k is also ergodic, then $P[N_k > m \mid N_0 = 0]\uparrow_k$ until the value $\sum_{m+1}^\infty e_n$ is reached.

§9.3. Monotone Chains in Continuous Time.

The bridge between discrete and continuous time permits many of the questions about the monotonicity of a chain in continuous time to be resolved at once. One extra little twist is present in continuous time that permits a class of continuous time chains to be monotone when the comparable discrete time chains are not. Because of this twist, as we will see, all birth-death processes are monotone, but all skip-free chains in discrete time are not.

A definition is needed.

Def. 9.3A. A Markov chain $N(t)$ in continuous time is "monotone" if $\mathfrak{p}(t) = [p_{mn}(t)]$ its transition probability matrix is monotone, i.e., if $\mathfrak{p}(t) \in \mathcal{M}_N$, all t.

Suppose that $N(t)$, governed by transition rates $\{\nu_{mn}\}$, is uniformizable (§2.1), so that $\mathfrak{p}(t) = \exp[-\nu t(I-\mathfrak{a}_\nu)]$. From (9.1.6), $N(t)$ is monotone when $\mathfrak{a}_\nu \in \mathcal{M}_N$, for any ν.

Consider a truncated birth-death process with reflecting states at $n = 0$, and $n = L$, governed by $\{\lambda_n, \mu_n\}$ $0 \leq n \leq L$, all positive except λ_L and μ_0 which are zero. Let $\nu > \mathrm{Max}(\lambda_j + \mu_j)$. Then $a_{\nu mn} = \nu_{mn}/\nu$, $m \neq n$; $a_{\nu mm} = 1 - \frac{\lambda_m + \mu_m}{\nu}$ and the tail sum matrix is given by

$$A_\nu = \begin{bmatrix} 1 & \frac{\lambda_0}{\nu} & 0 & 0 & \cdots & 0 \\ 1 & 1-\frac{\mu_1}{\nu} & \frac{\lambda_1}{\nu} & 0 & \cdots & 0 \\ 1 & 1 & 1-\frac{\mu_2}{\nu} & \frac{\lambda_2}{\nu} & \cdots & 0 \\ \cdot & \cdot & \cdot & \cdot & \cdot & \frac{\lambda_{L-1}}{\nu} \\ 1 & 1 & \cdot & \cdot & \cdot & 1-\frac{\mu_L}{\nu} \end{bmatrix}$$

Hence $a_{\nu mn}$ increases with m for n fixed provided that

$$\frac{\lambda_j + \mu_{j+1}}{\nu} < 1 \tag{9.3.1}$$

Clearly when L is finite or $N(t)$ is uniformizable, $N(t)$ is monotone. If $L = \infty$, and $N(t)$ is not uniformizable ($\lambda_j + \mu_j$ is unbounded) then it will be possible [50] to represent $N(t)$ as the limit of a sequence of truncated (finite) processes $N_j(t)$ each of which is monotone, with $\mathfrak{p}(t) = \lim \mathfrak{p}_j(t)$, all defined on $\mathscr{N} \times \mathscr{N}$ where $\mathscr{N} = \{0,1,2, \ldots \infty\}$, the convergence being for each component. But the limit of a sequence of monotone matrices is monotone. Hence all birth-death processes are monotone.

<u>Theorem 9.3B</u>. Every birth-death process is stochastically monotone.

This includes the class of birth-death processes for which the state $n = 0$ is absorbing since, as we have seen above, modification of monotone chains in discrete time by absorbing states does not affect the monotonicity.

Suppose now that one has a birth-death process commencing at $n = 0$. Then since $\mathfrak{p}(t)$ is monotone, and $\underline{p}(t) \succ \underline{p}(0)$ for every t no matter how small it follows that (Theorem 9.3F) the distribution vector $\underline{p}(t)$ increases stochastically for all t. Consequently, $E[N^k(t)]$ increases with t and $P[N(t) \geq m]$ increases with t for every state m. An application to reliability theory may be of interest.

<u>Example 9.3C</u>. Consider a system of M identical independent Markov components, K out of which must work for the system to work. Such a system is modeled by a truncated birth-death

process $N(t)$ with $\mu_m = m\mu$ and $\lambda_m = (M - m)\lambda$ when state m corresponds to the number of components working, and μ is the failure rate and λ the repair rate for a component. By the above it follows that $P[N(t) < m \mid N(0) = M]$ is monotonically increasing with t (since $N_1(t) = M - N(t)$ is also a birth-death process). We conclude that if the system starts in the perfect state and is unobserved, the probability that it is working decreases monotonically with time.

It is natural to ask for a characterization of those distributions $\underline{p}_0$ for which $\underline{p}(t)$ increases stochastically with time. The answer is contained in the following theorem [36], whose proof will not be given.

<u>Theorem 9.3D</u>. Let $N(t)$ be a birth-death process with $\lambda_0 > 0$. Let π_n be the potential coefficients (§3.3) $\pi_0 = 1$, $\pi_1 = \lambda_0/\mu_1$, $\pi_2 = \lambda_0\lambda_1/(\mu_1\mu_2)$, etc. Then $\underline{p}(t)$ increases stochastically with time iff $\underline{p}_0 = \underline{p}(0)$ is such that (p_n/π_n) decreases monotonically with n.

<u>Corollary 9.3E</u>. Let a system be modeled by an ergodic birth-death process for which $\{0,1,2, \dots k-1\}$ is the good set G. Then if $p_{0n} = e_n / \sum_0^{K-1} e_n$, $n \in G$, $p_{0n} = 0$ elsewhere, $\underline{p}(t)$ increases with t, i.e., $P[N(t) \in G \mid \underline{p}_0]$ decreases with t.

An extension of these methods permits one to discuss more general Markov chains in continuous time. Details may be found in [36].

The stochastic monotonicity of the sequence of probability vectors $\underline{p}(t)$ has a more general setting. The basic theorem is the following [36].

<u>Theorem 9.3F</u>. Let $N(t)$ be monotone and uniformizable, with

$P(t) = \exp[-\nu t(I - a_\nu)]$.

(a) If $\underline{p}(0) \prec \underline{p}(0)a_\nu$, then $\underline{p}(\tau) \prec \underline{p}(t + \tau)$ for all $t,\tau \geq 0$.

(b) If $\underline{p}(0) \prec \underline{p}(t)$ for all $t \geq 0$, then $\underline{p}(0) \prec \underline{p}(0)a_\nu$.

Proof: The following lemma, left to the reader is required.

Lemma 9.3C. If $\underline{q}_j \succ \underline{r}$, $j = 1,2, \ldots$ and $w_j \geq 0$, $\Sigma w_j = 1$, then $\sum_j w_j \underline{q}_j \succ \underline{r}$.

If $\alpha\underline{p} + (1 - \alpha)\underline{q} \succ \underline{q}$, for some α in $(0,1]$, then $\underline{p} \succ \underline{q}$.

Proof of (a): As above, $\underline{p}(0) \prec \underline{p}(0)a_\nu$ implies $\underline{p}(0) \prec \underline{p}(0)a_\nu^k$ for any integer k. Weighing these inequalities by Poisson probabilities $e^{-\nu t}(\nu t)^k/k!$, Lemma 9.3C gives

$$\underline{p}(0) = \underline{p}(0) \sum_0^\infty e^{-\nu t} \frac{(\nu t)^k}{k!} \prec \underline{p}_0 \sum_0^\infty e^{-\nu t} \frac{(\nu t)^k}{k!} a_\nu^k$$
$$= \underline{p}(0)P(t) = \underline{p}(t). \qquad (9.3.2)$$

Post-multiplying the outer hands of (9.3.1) by the monotone matrix $P(\tau)$, gives $\underline{p}(\tau) = \underline{p}(0)P(\tau) \prec \underline{p}(t)P(\tau) = \underline{p}(t + \tau)$.

Proof of (b): If $\underline{p}(0) \prec \underline{p}(t)$ then $\underline{p}(0) \prec \underline{p}(0)P(t) = \underline{p}(0)\exp(-\nu t[I - a_\nu])$, so that $\underline{p}(0) \prec e^{-\nu t}\,\underline{p}(0) + e^{-\nu t}\underline{p}(0)[\exp(\nu t a_\nu) - I]$. Now applying Lemma 9.3C gives $\underline{p}(0) \prec \underline{p}(0)[\exp(\nu t a_\nu) - I](e^{\nu t} - 1)^{-1}$. When $t \downarrow 0$, the right hand tends to $\underline{p}(0)a_\nu$, and the theorem is proved. □

§9.4. Other Monotone Processes in Continuous Time.

We have seen that the monotonicity of a implies that of $\exp\{-\nu t[I - a]\}$. It is also clear from the nature of stochastic monotonicity that the property is not intrinsic to

Markov chains, but is valid for any continuum process analogue of a monotone chain. Indeed, when such a process is expressible as the limit in law of a sequence of chains, where the grain of the lattice for the ordered chains goes to zero as the sequence index becomes infinite, the monotonicity property is preserved in the limit. For the process $X(t)$ one then has that $P[X(t) > x \mid X(0) = y]$ increases with y for every fixed x. The following processes are then seen to be monotone:

a. Markov Diffusion Process [8], [24].

These processes are the continuum analogues of birth-death processes, in that they are Markov and skip-free in both directions. (Their sample paths are continuous.) The class includes the Wiener processes (and their modifications by one or two reflecting or absorbing boundaries), and the Ornstein-Uhlenbeck process [8].

b. Homogeneous Markov Processes Modified by Boundaries.

A variety of processes of this type, described in §4.5, have applied and theoretical interest, amongst them the Lindley process (§4.5) and the Takacs process describing dams (§4.8). Such processes are of strong interest in sequential analysis.

One of the properties of a stochastically monotone process is that such a process is automatically "associated." Associated processes, introduced by Esary, Proschan and Walkup [13] lend themselves to important inequalities in reliability theory. These topics are treated at length in the forthcoming book of R. Barlow and F. Proschan [5].

References

[1] Arjas, E. and Nummelin, E., "Semi-Markov Processes and α-Invariant Distributions," Stoch. Proc. and their Appl. 6, 53-64, 1977.

[2] Arjas, E. and Speed, T. P., "Markov Chains with Replacement," Stoch. Proc. and their Appl. 3, 175-184, 1975.

[3] Bailey, N. T. J., MATHEMATICAL THEORY OF EPIDEMICS, Griffin, London, 1957.

[4] Barlow, R. E. and F. Proschan, MATHEMATICAL THEORY OF RELIABILITY, John Wiley, New York, 1965.

[5] Barlow, R. E. and F. Proschan, STATISTICAL THEORY OF RELIABILITY AND LIFE TESTING, Holt, Rinehart, Winston, New York, 1975.

[6] Berge, C., GRAPHS AND HYPERGRAPHS, North-Holland, 1973.

[7] Callaert, H. and J. Keilson, "On Exponential Ergodicity and Spectral Structure for Birth-death Processes," Stoch. Proc. Appl. 1, pp. 187-235, 1973.

[8] Cox, D. R. and H. D. Miller, THE THEORY OF STOCHASTIC PROCESSES, John Wiley & Sons, New York, 1965.

[9] Cramer, H. and M. R. Leadbetter, STATIONARY AND RELATED STOCHASTIC PROCESSES: SAMPLE FUNCTION PROPERTIES AND THEIR APPLICATIONS, John Wiley, New York, 1967.

[10] Daley, D. J., "Stochastically Monotone Markov Chains," Wahrscheinlichkeitstheorie verw. Geb. 10, pp. 305-317, (1968).

[11] Darroch, J. N., and E. Seneta, "On Quasi-Stationary Distributions in Absorbing Discrete-time Finite Markov Chains," J. Appl. Prob. 2, pp. 88-100, 1965.

[12] Debreu, G. and I. N. Herstein, "Nonnegative Square Matrices," Econometrica, 21, pp. 597-607, 1953.

[13] Esary, J. D., Proschan, F. and D. W. Walkup, "Association of Random Variables with Applications," Ann. Math. Stat. 38, pp. 1466-74, 1967.

[14] Feller, W., AN INTRODUCTION TO PROBABILITY THEORY AND ITS APPLICATIONS, Vol. 1, John Wiley, 1957.

[15] Feller, W., AN INTRODUCTION TO PROBABILITY THEORY AND ITS APPLICATIONS, Vol. II, John Wiley, New York, 1966.

[16] Hall, P., "On Measures of the Distance of a Mixture from its Parent Distribution," 1978 (to be published).

[17] Heyde, C. C., "Kurtosis and Departure from Normality," pp. 193-201, A MODERN COURSE ON STATISTICAL DISTRIBUTIONS IN SCIENTIFIC WORK, Proceedings of the NATO Advanced Study Institute, University of Calgary, 1974, edited by G. P. Patil, S. Kotz and J. K. Ord.

[18] Heyde, C. C. and Leslie, J. R., "On Moment Measures of Departure from the Normal and Exponential Laws," Stoch. Proc. and their Appl. 4, 317-328, 1976.

[19] Ibragimov, I. A., "On the Composition of Unimodal Distributions," Theory Prob. Appl. 1, 1956.

[20] Kalmykov, G. I., "On the Partial Ordering of One-Dimensional Markov Processes," Theor. Prob. Appl. 7, pp. 456-459, 1962.

[21] Karlin, S., TOTAL POSITIVITY, Stanford University Press, Stanford, Ca. 1968.

[22] Karlin, S., Proschan, F. and R. E. Barlow, "Moment Inequalities of Polya Frequency Functions," Pac. J. Math, Vol. II, pp. 1023-1033, 1969.

[23] Keilson, J., GREEN'S FUNCTION METHODS IN PROBABILITY THEORY, Charles Griffin, 1965.

[24] Keilson, J., "A Review of Transient Behavior in Regular Diffusion and Birth-death Processes, Part II," J. Appl. Prob. 2, pp. 405-428, 1965.

[25] Keilson, J., "The Ergodic Queue Length Distribution for Queueing Systems with Finite Capacity," J. Roy. Stat. Soc. B, Vol. 28, No. 1, pp. 190-201, 1966.

[26] Keilson, J., "A Limit Theorem for Passage Times in Ergodic Regenerative Processes," Ann. Math. Stat. 37, pp. 866-870, 1966.

[27] Keilson, J., "A Technique for Discussing the Passage Time Distribution for Stable Distributions," J. Roy. Stat. Soc. B, Vol. 28, No. 3, pp. 477-486, 1966.

[28] Keilson, J., "Log-concavity and Log-convexity in Passage Time Densities of Diffusion and Birth-Death Processes," J. Appl. Prob. 8, pp. 391-398. 1971.

[29] Keilson, J., "A Threshold for Log-concavity for Probability Generating Functions and Associated Moment Inequalities," Ann. Math. Stat. 43, pp. 1702-1708, 1972.

[30] Keilson, J., "Sojourn Times, Exit Times, and Jitter in Multivariate Markov Processes," Adv. Appl. Prob. 6, pp. 747-756, 1974.

[31] Keilson, J., "Stochastic Order in Renewal Theory and Time-Reversible Chains," Department of Statistics, Stanford University Technical Report No. 82, 1974.

[32] Keilson, J., "Systmes of Independent Markov Components and Their Transient Behavior," Reliability and Fault Tree Analysis, SIAM, Philadelphia, 1975, pp. 351-364.

[33] Keilson, J., "Exponential Spectra as a Tool for the Study of Server Systems," J. of App. Prob. 15, pp. 162-170, 1978.

[34] Keilson, J. and H. Gerber, "Some Results for Discrete Unimodality," Journal Amer. Stat. Assn. 66, pp. 386-389, 1971.

[35] Keilson, J. and Graves, S., "Methodology for Studying the Dynamics of Extended Logistic Systems (to appear in Naval Research Logistics Quarterly).

[36] Keilson, J., and A. Kester, "Monotone Matrices and Monotone Markov Processes," Stoch. Proc. and their Appl., Vol. 5, No. 3, pp. 231-241, 1977.

[37] Keilson, J. and Kester, A., "Unimodality Preservation in Markov Chains," Stoch. Proc. and Their Appl. 7, 179-190.

[38] Keilson, J. and F. W. Steutel, "Families of Infinitely Divisible Distributions Closed Under Mixing and Convolution," Ann. Math. Stat. 43, pp. 242-250, 1972.

[39] Keilson, J. and F. W. Steutel, "Mixtures of Distributions, Moment Inequalities and Measures of Exponentiality and Normality," Annals of Probability 1, 1974.

[40] Keilson, J. and S. Subba Rao, "A Process with Chain Dependent Growth Rate," J. Appl. Prob. 7, pp. 699-711, 1970.

[41] Keilson, J. and Syski, R., "Compensation Measures in the Theory of Markov Chains," Stoch. Proc. and Their Appl. 2, 1974, 59-72.

[42] Keilson, J. and D. M. G. Wishart, "Boundary Problems for Additive Processes Defined on a Finite Markov Chain," Proc. Camb. Phil. Soc. G1, pp. 173-190, 1965.

[43] Kellogg, O. D., FOUNDATIONS OF POTENTIAL THEORY, Dover Publications, New York, 1953.

[44] Kelly, F. P., "Networks of Queues with Customers of Different Types," J. of App. Prob. 12, 1975, 542-554.

[45] Kelly, F. P., "Stochastic Population Models in Genetics," J. Appl. Prob. 13 (1), 1976, 127-131.

[46] Kemeny, J. G. and J. L. Snell, FINITE MARKOV CHAINS, Van Nostrand, Princeton, N. J., 1960.

[47] Kingman, J. F. C., "A Convexity Property of Positive Matrices," Quart. J. Math. Oxford 12, pp. 283-284, 1961.

[48] Kingman, J. F. C., "Markov Population Processes," J. J. Appl. Prob. 6, 1969, 1-18.

[49] Kirstein, B. M., "Monotonicity and Comparability of Time-Homogeneous Markov Processes with Discrete State Space," Math. Operationsforschung u. Stat. 7, 1976, 151-168.

[50] Ledermann, W. and G. E. H. Reuter, "Spectral Theory for the Differential Equations of Simple Birth-and-Death-Processes," Philos. Trans. Royal Soc. A. 246, pp. 321-369, 1954.

[51] Lehmann, E. L., "Some Concepts of Dependence," Ann. Math. Stat. 37, 1966, 1137-1153.

[52] Lindley, D. V., "The Theory of Queues with a Single Server," Proc. Camb. Phil. Soc. 48, pp. 277-289, 1952.

[53] Mitrinovic, D. S., ANALYTIC INEQUALITIES, Springer, Berlin, 1970.

[54] Nummelin, E., "Limit Theorems for α-Recurrent Semi-Markov Processes," Adv. in App. Prob. 8, 1976, 531-547.

[55] Nummelin, E., "On the Concepts of α-Transience for Markov Renewal Processes," Stoch. Proc. and their Appl. 5, 1977, 1-19.

[56] O'Brien, G., "Comparison Theorems for Stochastic Processes." Ph.D. Thesis, Dartmouth College, 1971.

[57] O'Brien, G., "A Note on Comparisons of Markov Processes," Ann. Math. Stat. 43, pp. 365-368, 1972.

[58] Reich, E., Departure Processes, CONGESTION THEORY (Edited by W. L. Smith and W. E. Wilkinson), Univ. N. Carolina Press, 1965.

[59] Seneta, E. and D. Vere-Jones, "On Quasi-stationary Distributions in Discrete Time Markov Chains with a Demumerable Infinity of States," J. Appl. Prob. 3, pp. 403-434, 1966.

[60] Spitzer, F., PRINCIPLES OF RANDOM WALK, D. Van Nostrand, Princeton, 1964.

[61] Syski, R., "Potential Theory for Markov Chains" in PROBABILISTIC METHODS IN APPLIED MATHEMATICS V3, Academic Press, pp. 213-276, New York, 1973.

[62] Syski, R., "Perturbation Models," Stoch. Proc. and their Appl. 5, 1977, 93-130.

[63] Syski, R., "Ergodic Potential," Stoch. Proc. and their Appl. 7, 1978, 311-336.

[64] Veinott, A. F., "Optimal Policy in a Dynamic, Single Product, Nonstationary Inventory Model with Several Demand Classes," Oprns. Res. 13, pp. 761-778, 1965.

[65] Whitt, W. and Sonderman, D., "Comparing Continuous Time Markov Chains," Preprint, School of Organization and Management, Yale University, 1976.

[66] Widder, D. V., THE LAPLACE TRANSFORM, Princeton University Press, Princeton, 1946.

Index

Aggregation of states, 41

Aperiodicity in Markov chains, 16

Arjas, E., 92, 105

Artin, Emil, 66

Backward equation, 21

Barlow, R. E., 69, 176

Birth-death process, 25, 57

homogeneous, 25

ergodic, 27

Boundary state, 47, 84

Censored transition, 39

Chain [cf. Markov chains]

Chapman-Kolmogorov equation, 16

Compensation measure, 50, 51

Compensation method, 47

Complete monotonicity, 34, 63

in time-reversible process, 67

Daley, D. J., 164

Detailed balance

in discrete time, 18

in continuous time, 27

DFR class, 57, 74

Dominate, 164

Dual of transition matrix, 18

Ergodicity for Markov chains

discrete time, 16,

continuous time, 25, 26

Esary, J. D., 176

Exit time [Cf. sojourn time], 68, 76

Ergodic, 76, 88, 102, 130

mean, 97

quasi-stationary, 76, 88, 90, 130

Failure rate [Cf. hazard rate], 74

Failure time

distribution, 88

mean, 112

Feller, W., 21, 24, 34, 66, 136, 155

Flow rate,

ergodic, 86

Forward equation, 21

Fundamental matrix

discrete time, 106

continuous time, 107

Graves, S., 30

Green measure

ergodic, 48

time dependent, 48

Green potential, 44, 84

Hall, P., 155

Hazard rate, 74

Heyde, C. C., 155

Ibragimov, I. A., 64

IFR class, 57, 76

Increment measure, 47

Irreducibility in discrete time Markov chains, 16, 17

Jitter, 102, 150
  factor, 150

Kalmykov, G. I., 164

Karlin, S., 65, 67

Keilson, J., 30, 55, 60, 68, 69, 105, 164

Kelly, F. P., 30

Kester, A., 164

Kirstein, B. H., 164

Kingman, J.F.C., 30, 66

Lehmann, E. L., 164

Leslie, J. R., 164

Lindley process, 49

Log-concavity, 57, 64

Log-convexity, 57, 64

Markov chains
  in discrete time, 16
  in continuous time, 20
  monotone in continuous time, 171
  monotone in discrete time, 168
  skip-free negative homogeneous, 52
  skip-free positive, 54
  transient, 31, 48
  lossy, 31, 44

Markov diffusion process, 175

Mass
  compensation, 53

Mass
  neutralizing, 51
  replacement, 51

Mintrinovic, D. S., 65

Monotone chain, 164

Monotone Markov matrix, 164

Multicycle time, 156

Nummelin, E., 92

O'Brien, G., 164

Ornstein-Uhlenbeck diffusion process, 102

Passage time, 76, 122
  first, 16, 77
  mean, 81
  moment, 61

Passage time density, 57, 76
  upward, 57

Path, 35
  independence, 31
  product, 35

Perron-Romanovsky-Frobenius theorem, 18

PF_∞, 63

Poisson queue, 25

Polya-frequency function, 57, 63

Potential, 35, 43
  homogeneous, 51
  coefficient, 31, 36, 61

Principal decay rate, 91

Probability vector, 15

Process
  Associated, 175
  ergodic and regenerative, 12
  J-, 96
  Markov, 15
  monotone Markov, 164
  neutralizing, 51
  parent, 158
  performance, 119
  temporally homogeneous, 15
  uniformizable, 22
Proschan, F., 69, 176
Random Poisson transition epoch, 20
Random walk, 24
Recurrent state, 16
  positive, 16
  null, 16
Regenerative state, 133
Relaxation time, 161
Replacement process, 41
  ergodic distribution for, 45
Reversibility in time, 18
  discrete time, 19
  continuous time, 26, 38
Ross, S., 149
Ruin probability, 79, 84
  for systems with independent Markov components, 112
Smith, D. R., 161
Sojourn time, 76, 130
  ergodic, 76, 88
  expected, 95
  mean, 97
  density, 67
Sonderman, D., 164
Spectral representation of the time dependent behavior, 32 35
Speed, T. P., 105
Spitzer, F., 55
State made absorbing, 40
State probability vector, 21
State space, 15
Stationarity in discrete Markov chains, 17
Stochastic ordering, 99, 100, 165
Stochastically monotone, 164
Strongly stable chain, 151
Substochastic matrix, 37
  strictly, 44
Survival function, 92
Syski, R., 105
Tail sum vector, 165
Time-reversible chain [Cf. reversibility], 18
Total positivity, 65
Transient state, 16
Transition kernel, 49

Transition probability matrix, 15, 20

Tree process, 28, 29

Uniformizable chain, 22, 24

Uniformization procedure between discrete and continuous time Markov chains, 20

Unimodality

for continuous distribution, 63

for discrete distribution, 70

lattice, 70

strong, for continuous distribution, 57, 63

strong, for discrete distribution, 70

Veinott, A. F., 164

Walkup, D. W., 175

Whitt, W., 164

Widder, D. V., 66

Wiener process, 175

Applied Mathematical Sciences

Vol. 1 F. John
Partial Differential Equations
Third edition
ISBN 0-387-90327-5

Vol. 2 L. Sirovich
Techniques of Asymptotic Analysis
ISBN 0-387-90022-5

Vol. 3 J. Hale
Theory of Functional Differential Equations
ISBN 0-387-90203-1 (cloth)

Vol. 4 J. K. Percus
Combinational Methods
ISBN 0-387-90027-6

Vol. 5 R. von Mises and K. O. Friedrichs
Fluid Dynamics
ISBN 0-387-90028-4

Vol. 6 W. Freiberger and U. Grenander
A Short Course in Computational Probability and Statistics
ISBN 0-387-90029-2

Vol. 7 A. C. Pipkin
Lectures on Viscoelasticity Theory
ISBN 0-387-90030-6

Vol. 8 G. E. O. Giacaglia
Perturbation Methods in Non-Linear Systems
ISBN 0-387-90054-3

Vol. 9 K. O. Friedrichs
Spectral Theory of Operators in Hilbert Space
ISBN 0-387-90076-4

Vol. 10 A. H. Stroud
Numerical Quadrature and Solution of Ordinary Differential Equations
ISBN 0-387-90100-0

Vol. 11 W. A. Wolovich
Linear Multivariable Systems
ISBN 0-387-90101-9

Vol. 12 L. D. Berkovitz
Optimal Control Theory
ISBN 0-387-90106-X

Vol. 13 G. W. Bluman and J. D. Cole
Similarity Methods for Differential Equations
ISBN 0-387-90107-8

Vol. 14 T. Yoshizawa
Stability Theory and the Existence of Periodic Solutions and Almost Periodic Solutions
ISBN 0-387-90112-4

Vol. 15 M. Braun
Differential Equations and Their Applications
ISBN 0-387-90114-0

Vol. 16 S. Lefschetz
Applications of Algebraic Topology
ISBN 0-387-90137-X

Vol. 17 L. Collatz and W. Wetterling
Optimization Problems
ISBN 0-387-90143-4

Vol. 18 U. Grenander
Pattern Synthesis
Lectures in Pattern Theory Vol. I
ISBN 0-387-90174-4

Vol. 19 J. E. Marsden and M. McCracken
The Hopf Bifurcation and Its Applications
ISBN 0-387-90200-7

Vol. 20 R. D. Driver
Ordinary and Delay Differential Equations
ISBN 0-387-90231-7

Vol. 21 R. Courant and K. O. Friedrichs
Supersonic Flow and Shock Waves
ISBN 0-387-90232-5 (cloth)

Vol. 22 N. Rouche, P. Habets, and M. Laloy
Stability Theory by Liapunov's Direct Method
ISBN 0-387-90258-9

Vol. 23 J. Lamperti
Stochastic Processes
A Survey of the Mathematical Theory
ISBN 0-387-90275-9

Vol. 24 U. Grenander
Pattern Analysis
Lectures in Pattern Theory Vol. II
ISBN 0-387-90310-0

Vol. 25 B. Davies
Integral Transforms and Their Applications
ISBN 0-387-90313-5

Vol. 26 H. J. Kushner and D. S. Clark
Stochastic Approximation Methods for Constrained and Unconstrained Systems
ISBN 0-387-90341-0

Vol. 27 C. de Boor
A Practical Guide to Splines
ISBN 0-387-90356-9

Vol. 28 J. Keilson
Markov Chain Models
Rarity and Exponentiality
ISBN 0-387-90405-0